Silvicultural Systems

Silvicultural Systems

JOHN D. MATTHEWS
Emeritus Professor of Forestry,
University of Aberdeen

CLARENDON PRESS · OXFORD

Oxford University Press, Walton Street, Oxford OX2 6DP
Oxford New York Toronto
Delhi Bombay Calcutta Madras Karachi
Petaling Jaya Singapore Hong Kong Tokyo
Nairobi Das es Salaam Cape Town
Melbourne Auckland
and associated companies in
Berlin Ibadan

Oxford is a trade mark of Oxford University Press

Published in the United States
by Oxford University Press, New York

First published 1989
First published in paperback (with corrections) 1991

British Library Cataloguing in Publication Data
Matthews, John D.
Silvicultural systems.
1. Trees. Cultivation
I. Title
634.9'5
ISBN 0–19–854670–X (Pbk)

Library of Congress Cataloging in Publication Data
Matthews, John D.
Silvicultural systems / John D. Matthews.
p. cm. Bibliography: p.
Includes index.
1. Silvicultural systems. I. Title.
SD392.M38   1989   634.9'5—dc19
89–3145   CIP
ISBN 0–19–854670–X (Pbk)

Printed by St Edmundsbury Press
Bury St Edmunds, Suffolk

Preface

In Europe, until the late eighteenth century, timber was one of the most basic requirements of human existence. Essential to most forms of manufacture and industry, almost all types of building, the production of most implements and machines, and to almost every aspect of domestic and agricultural life, timber and its allied products played a central role in the economy of Europe until the time when its dominance was slowly eroded by the use and availability of new fuels and materials. 'To use a modern analogy . . . it was as if the economic and technological functions of oil, steel, and man-made plastics were combined in one type of material source' (Sharp 1975).

The forests from which the timber came also had important functions. In many parts of Europe they were valued primarily for the direct protection they gave to water supplies, and in mountainous areas the maintenance of forests provided the only cheap and practical means of preserving the soil from erosion. The influence that forests exert by providing shelter from strong and drying winds was of great importance in growing some crops and husbanding livestock. Forests also often contained a valuable store of wild animals and plants used as food and, in the case of some plants, as medicine.

For a long period the favourable climate of the region enabled the central European forest to regenerate itself, even in the face of human use, but increases in population during the sixteenth and especially in the eighteenth century eventually led to fears of a wood shortage and hence disaster due to lack of domestic fuel and adverse effects on industry and trade. Then it became necessary to plan the management of forests, raise the yield of timber and other products, control the rate and type of utilization and conserve the area of forest. So it was that permanent forest administrations, staffed by professional foresters first appeared in Europe, and forest science began to systematize and improve the techniques arrived at by accumulated practical experience.

In the countries of the developing world today the demand for timber rises as populations increase but so also does the demand for food. The demand for regular supplies of clean water for people, fisheries, agriculture, and industry is also rising but forests are being destroyed on water catchments, and soil erosion is damaging ever larger areas of the mountainous regions that supply water. These problems are very old but today their enormous scale causes great concern throughout the world (Brundtland 1987).

As Troup (1928) put it 60 years ago ... 'the depletion of natural forests that has taken place during the past hundred years in many parts of the world gives genuine cause for alarm and steps have been taken in many countries to conserve and protect at least some small proportion of their original forest area. This, however, is only an initial step; if the problem of future timber supplies is to be solved, it is also of the utmost importance that the reduced forest area now available should be treated in such a way as to produce the highest possible sustained yield of suitable timber compatible with economic and other considerations. Certain European countries were faced with this problem centuries ago, and as a result of long experience have evolved methods of treatment, termed silvicultural systems, which are an object lesson to the whole world. The detailed study of these systems under as many different conditions as possible is the only means of acquiring that special knowledge which will lead to their intelligent application in practice.'

Although every endeavour has been made to describe the various systems as they are actually practised today, the reader is asked to regard this book not as a means of acquiring a complete knowledge of silvicultural systems but rather as a guide towards the practical application of these systems in the forest.

Low Row
October 1988.

J.D.M.

Preface to the paperback edition

In this paperback edition some minor changes have been made in the light of papers presented at the meeting held by the Institute of Chartered Foresters to discuss silvicultural systems in Britain (Gordon 1991).

Heswall
April 1991

J.D.M.

Reference

Gordon, P. (ed.) (1991). *Silvicultural systems*. Institute of Chartered Foresters, Edinburgh.

Acknowledgements

Several organizations and many people have helped me during the preparation of this book. I am very grateful to the Scottish Forestry Trust and the Leverhulme Trust for awarding grants toward the cost of preparing the text, diagrams, and figures. Miss Jacqueline Kelly B.Sc. (Forestry) provided excellent research assistance. Mr Peter Seal and Mr Marcus Byron drew the diagrams.

Permission to reproduce photographs was given by:

British Forestry Commission, Edinburgh   Fig. 2
Paul Haupt, Bern, Switzerland   Fig. 27
Hedeselskabet, Denmark   Fig. 1
Forestry Agency, Tokyo, Japan   Figs. 3, 6
Forest Research Institute, Rotorua, New Zealand   Fig. 4
Ms Jacqueline Kelly   Figs. 17, 23, 24, 29
The late W. M. McNeill   Figs. 5, 10, 11, 14
Mr Ken Taylor   Fig. 9

I acknowledge with great pleasure the help provided by the following people: Professor George Holmes, Professor Hugh Miller, Mr David Seal, and Mr Jack Evans for their support and many useful discussions. Mr Donald Cumming and Mr Michael Philip for information about tropical forestry. Mr Nigel Hector and Mr Edward Liddon for the account of basket willows in Somerset. Dr John Blyth, Mr Roy Dyer, Mr J. E. Garfitt, Dr Douglas Malcolm, Mr S. A. Neustein, Mr M. G. Reade, and Mr Jack Watson for information about the application of the group selection system in Britain. Mr Robert Hide and Mr George Stevenson for discussions on the silviculture of broad-leaved species in Britain. Professor J. Hüss, Universität Freiburg; Dr Robert Holzapfl, Leitender Forstdirektor, Ottobrunn-Riemerling, West Germany; Professor H. Thomasius, Technische Universität Dresden; and Dr H. Oswald, Centre National de Recherches Forestières, Champenoux, France for valuable information about silvicultural systems in Europe. Mr Keith Illingworth, Mr Christopher Heaman, Mr Steven Northway, and Mr Glen Dunsworth for discussions about silviculture in British Columbia.

The first edition of 'Silvicultural systems' was written by the late Professor R. S. Troup FRS in 1926 and revised by his colleague Dr Eustace Jones in 1952. Advances in forest genetics, tree physiology, and forest ecology have transformed silviculture but the principles expounded by these two scholars have often survived intact. Thus it gives me great pleasure to acknowledge their influence on the form and content of this book.

Lastly I wish to thank Dr Julian Evans and the Oxford University Press for suggesting that I should prepare this new text on 'Silvicultural systems'.

Contents

animal damage   Cutting sections   Form and arrangement
of coupes in hilly country   Design of coupes in relation to
the landscape: Clear cutting with artificial regeneration
Treatment of slash   Ground cultivation and drainage
Weed control   Supplementary nutrition   Special
techniques for artificial regeneration: Artificial regeneration
by direct seeding: Artificial regeneration with the aid of
field crops: Clear cutting with natural regeneration
Regeneration from seed already on the area   Regeneration
from seed disseminated from outside   Patch felling
Strip-like clear cuttings   Progressive fellings   Fellings in
alternate strips: Advantages and disadvantages of the clear
cutting system: Application in practice

To Marjorie, George, and Peter

PART I

The theoretical background to the systems

1

Introduction

A silvicultural system may be defined as the process by which the crops constituting a forest are tended, removed, and replaced by new crops, resulting in the production of stands of distinctive form (Fig. 1). In this definition 'tending' refers mainly to thinning operations done in immature crops, in so far as these operations affect the state of the crop and soil at the time of regeneration.

A silvicultural system embodies three main ideas:

(1) the method of regeneration of the individual crops constituting the forest;
(2) the form of crop produced; and
(3) the orderly arrangement of the crops over the whole forest, with special reference to silvicultural and protective considerations and efficient harvesting of produce.

The purpose of this book is to describe the silvicultural systems in common use. We shall be concerned with the techniques of each system, including the conduct of felling, tending, and regeneration. We will also examine the application of each system and this is closely connected with the requirements of forest management. Forest management teaches us the great economic advantages of regular sustained yields over intermittent or spasmodic yields, and the ideal of the normal growing stock has been created with the object of ensuring future sustained yields. This must be kept in mind when applying a silvicultural system to any regulated scheme of management which aims at this ideal. So, we will consider the general framework of the scheme including the division of the forest, the allotment of areas for felling, and also yield regulation as they affect silvicultural, protection, and harvesting produce.

Forests have productive, protective, and social functions. The protective functions of some forests are of equal or greater importance than their productive functions and these 'protection forests' are managed so as to assist in controlling soil erosion and avalanches, protect water supplies, and provide habitat for wild plants and animals. Such forests may also be

Figure 1.   Alternate strip system with artificial regeneration on level ground, showing different stages of regeneration. Norway spruce, Jutland, Denmark.

essential parts of well-loved landscapes. The social functions of forests include providing employment and facilities for field sports and recreation. The relative values of the various silvicultural systems in satisfying the productive, protective, and social functions of managed forests will be described and assessed.

The study of silvicultural systems presupposes a knowledge of the foundations on which silviculture is based. Genetics, physiology, and ecology are fully established as foundation subjects and they have been joined by aspects of physics, mechanics, the social sciences, and many more. To avoid repetition as each silvicultural system is described, several topics fundamental to many and often all of them are introduced in Chapters 2, 3, 4 and 5. Comprehensive treatment of each topic is not attempted; the aim is to set scenes so that the accounts of the systems are clear and concise.

Various classifications of silvicultural systems have been proposed. If we take into consideration all the variations of treatment that are in use at present or have been practised in the past, there is scarcely any limit to the number of these variations. Thus it is necessary to group the various forms of treatment into major systems and devise a general classification of these.

Opinions differ considerably as to how such a classification should be made, but for our purposes the following will be suitable:

High forest systems. Crops normally of seedling origin.
  Felling and regeneration for the time being concentrated on part of the forest area only:
    Old crop cleared by a single felling; resulting crop even-aged—*Clear cutting systems*
    Systems of successive regeneration fellings. Old crop cleared by two or more successive fellings; resulting crop more or less even-aged or somewhat uneven-aged:
      Regeneration fellings distributed over whole compartments or sub-compartments:
        Opening of canopy even; young crops more or less even-aged and uniform—*Uniform systems*
        Opening of canopy by scattered gaps; young crop more or less even-aged—*Group system*
        Opening of canopy irregular and gradual; young crop somewhat uneven-aged—*Irregular shelterwood system*
      Regeneration fellings confined to certain portions of compartments or sub-compartments at a time:
        Fellings in strips—*Strip systems*
        Fellings beginning in internal lines and advancing outwards in wedge formation—*Wedge system*
  Felling and regeneration distributed continuously over the whole area; crop wholly uneven-aged (irregular)—*Selection systems*
    Accessory systems arising out of other systems:
      Form of forest produced by introducing a young crop beneath an existing immature one—*Two-storied high forest*
      Form of forest produced by retaining certain trees of the old crop after regeneration is completed—*High forest with reserves*
Coppice systems. Crops, in part at least, originating from stool shoots (coppice) or by other vegetative means:
    Crop consisting entirely of vegetative shoots:
      Crop removed by clear felling; even-aged—*Coppice system*
      Only a portion of the shoots cut at each felling; crop uneven-aged—*Coppice selection system*
    Crop consisting partly of vegetative shoots, partly of trees generally of seedling origin—*Coppice with standards system*

Shelterwood systems is a general term which comprises systems of successive regeneration fellings and the selection systems (see pages 90 and 163). The term 'even-aged' is synonymous with 'uniform' or 'regular' and 'uneven-aged' is synonymous with 'irregular' (see page 15). The terms 'stand' and 'crop' are both used to denote silvicultural or management units that are homogeneous in one or several respects.

2

Forest ecology and genetics

The forest ecosystem

A forest ecosystem is the product of the climate, geology, terrain and soil, and of the trees, shrubs, animals, fungi, and other organisms living and interacting together on the site. The canopy of leaves and branches supported by the tree stems intercepts incoming radiation and precipitation in the form of rain, snow, or mist. Some radiation is reflected back to the atmosphere and some precipitation is evaporated from the surface of the crowns but the larger proportion of these enter the forest ecosystem together with nutrients borne by rain, dust, and aerosols. The canopy also affects the flow of air over and through the upper parts of the forest, making it turbulent.

When the microclimate within a forest stand is compared with that of a nearby site without trees, the former is more equable. The extremes of air and soil temperatures are lessened. The amount of precipitation, that penetrates the canopy as throughfall and stem flow is reduced and the impact on the soil surface is lessened. Wind speeds in the lower parts of the forest are one-quarter to one-half of those in the open, the humidity of the air is higher than outside, and the amount of light reaching the ground when the full canopy is present is between 1 and 15 per cent of incident light.

The influence of the canopy on the soil is exercised through its role in the cycle of nutrients. As they die and fall the leaves, twigs, flowers, fruits, and fragments of bark become the layer of litter covering the ground, which as it decomposes to release nutrients and form humus, helps to maintain the vigour and health of the tree crop and the physical and chemical properties of the soil. The litter layer is inhabited by microfauna and microflora, including mycorrhizae. It is also explored by very many fine roots absorbing water and nutrients. If the layers of decomposing litter are developing normally they improve the infiltration of water into the soil, so that surface run-off and erosion are rare under the canopy of a forest and stream flow from the site is regulated in its amount and quality.

The structural roots anchor the trees to the soil and the channels created by their elongation, radial thickening, and eventual death improve the physical structure of the soil, particularly its permeability. Those roots which reach the lower horizons bring up water and also nutrients released by weathering. These are translocated to the leaves which eventually fall to become part of the litter layer; the nutrients subsequently released enrich the upper horizons of the soil. The flows of water, nutrients, and energy through a mature forest ecosystem are combined and regulated by its living and non-living parts. Losses from the ecosystem are small so biomass and nutrients accumulate within it.

When the trees are clear felled, the overhead and side shelter provided by the canopy is lost and the microclimate of the site becomes similar to that of a nearby site without trees. The range of temperature widens and the desiccative effects of wind have full play. Several processes of the forest ecosystem cease to operate:

1. The hydrological cycle is broken, interception with evaporation and transpiration cease, and the water table rises. Wet soils become wetter and dry soils dryer. Surface run-off also increases.
2. The nutrient cycle is broken. The litter and humus layer are exposed to sun, rain, and wind so breakdown is rapid. There may be some loss of nutrients from the site or they may be taken up by colonizing vegetation.
3. The annual increment of biomass virtually ceases and some of the accumulated capital is removed from the site. If only the tree stems are harvested, a small part of the biomass and nutrients is lost and will eventually be replaced by inputs of atmospheric nutrients and soil weathering. If the foliage, stems, and roots are harvested the nutrient capital of a site of low fertility may be significantly reduced.

Stated in ecological terms (Whitehead 1982), the objects of silvicultural systems are first, to harvest an appropriate part of the biomass so that the productivity of the ecosystem remains undiminished in the long term; and second, to adjust the canopy and control the felling and extraction of produce, so that the microclimate of the site and condition of the soil are favourable for regeneration with trees of suitable species. In these circumstances the disruption of the forest ecosystem is minor and its effects are short-lived.

If the forest ecosystem is treated without due care and the site cannot readily be restocked, the disturbance caused by felling and harvesting is more severe and long-lived. The site may have become less productive. The most common symptoms of site deterioration are:

(1) soil compaction and reduction of rootable volume;
(2) excessive wetness or dryness of the surface and upper horizons of the soil;

(3) loss of soil horizons due to surface run-off and erosion;
(4) inactivation or destruction of the humus layer;
(5) rank growth of weeds which compete with young trees;
(6) accumulation of damaging soil-borne fungi; and
(7) decreased or irregular growth of the new crop.

The repeated passage of logs and harvesting equipment over the soil can compact the surface and upper horizons, reducing permeability and microscopic pore space which in turn increases bulk density and impairs infiltration (Conway 1982). These effects are greatest on soils containing much silt or clay, especially when they are wet, and least on sandy soils. They can be greatly reduced by orderly felling of the trees by trained fellers, extraction along prepared routes, and speedy restoration of the physical condition of the soil and the drainage system of the site. Timing the harvesting to coincide with dry periods when the ground is firmer than usual is another precaution taken to avoid damaging soils with poor load-bearing capacity. Building forest roads, which involve considerable cut and fill, can be a potent source of site deterioration unless care is taken in their design and construction (see page 56).

Three common causes of site deterioration are:

(1) loss of the litter layer by repeated burning at short intervals;
(2) physical removal of the litter layer for fuel and other purposes; and
(3) excessive grazing by domestic or wild animals.

Where these practices involve rights of user they can be difficult to correct. In the case of grazing rights, definite areas may have to be closed so that they can be regenerated. Clear cutting and other systems with short regeneration periods are preferable to those with long periods (see pages 108 and 143).

Composition and structure of forest stands

Natural forest ecosystems vary greatly in their composition and structure. They may be composed of one or several tree species which are similar in age or size, or they may differ in these respects. Two important aspects of forest ecology requiring discussion are whether managed forests should comprise stands that are pure or mixed in species composition and whether they should be regular or irregular in structure. Figures 2 and 3 depict these forms of stand.

When modern forestry began in central Europe during the eighteenth

Figure 2.   Pure regular high forest of Western hemlock, aged 24 years, after second thinning. The stems of some trees have been pruned for purposes of mensuration. Gwydyr forest, Wales.

century, there was a strong tendency to adopt rigid methods of silviculture with regular crops of one species and attendant rigid methods of management. These were appropriate to the time. The early foresters had to re-establish and expand the depleted forest resource and introduce methods of silviculture that met, on a sustained yield basis, very large demands for timber. Once that process was under way and it became apparent that there were conditions of climate, soil, and terrain unsuited to the growth of pure, regular stands of some species, silviculture and management became less stereotyped.

Before the close of the nineteenth century many European foresters, led

Figure 3. Irregular high forest of hinoki (*Chamaecyparis obtusa*). Some maple (*Acer* sp.) is also present. Japan.

by Gayer (1880, 1886), were moving steadily towards more flexible methods of silviculture and management and hence more irregular forms of forest stand (Troup 1952, p. 194). Intimately connected with the increasing belief in irregularity was a growing conviction that mixtures are desirable. The advocates of the 'natural' approach to forestry require that silviculture should be based on the ecology of natural indigenous forests. In consequence the emphasis in silviculture is on adjustment to rather than radical changes in natural forms of forest ecosystems (see, for example, Susmel 1986).

Contrasting with this 'natural' forestry is the modern approach to the silviculture of pure, regular stands. In many countries, new forests are usually established on soils formerly in agriculture and not bearing trees, or they replace a forest type which, because of its composition or condition, is not suitable as a source of useful timber. The basic concept of the forest as an ecosystem is not forgotten; rather the emphasis has shifted to developing techniques that increase production of a specified raw material for industry.

Several conditions must be satisfied if pure, regular stands are to become viable forest ecosystems. The physical condition and fertility of the soil must be maintained, and if possible improved. The species must be well adapted to the climate and to growth in pure, regular stands so that the trees remain vigorous and do not become susceptible to attack by diseases and pests. The trees must also be able to respond to the silvicultural treatments applied (see page 71).

Stone (1975) has pointed out that the reciprocal influences of soil on forest and of forest on soil are not easily disentangled, partly because of the long periods over which these influences act. Presenting a unified account of the influence of forests on soil is also made difficult 'by the burden of supposition and lack of rigour that dominates many accounts of soil changes in the forestry and ecological literature'. The response of the soil to forest cover can be examined from the viewpoints of nutrient cycling, soil genesis and classification, and of the relatively short term changes in soil properties or soil productivity. It is the last of these that is considered here.

Much of the interest in adverse soil changes stems from the concerns of European silviculturists faced with managing great areas of planted Norway spruce and other conifers that had replaced degraded woodlands, heaths, pastures and unprofitable agriculture during the nineteenth century. The severe and often prolonged checking of young, pure conifer plantations has been the most serious and the most elusive of the troubles affecting them. By the late nineteenth century in Saxony and Bavaria, there were many young second and third generation crops of Norway spruce following broad-leaved crops which were unhealthy and growing very slowly in places where, it was alleged, their predecessors had grown well. Weidemann (1923, 1924), who studied the problem in lowland Saxony, found every graduation between crops that had merely grown considerably more slowly than their neighbours, often because they had passed through periods of very slow growth in their youth, to crops that had turned yellow and practically ceased to grow. There was also every graduation between crops that had escaped from check unaided after a few years and those that remained in check indefinitely.

Weidemann pointed out that check occurred in areas of low rainfall and that periods of check began in years with summer drought. He also found that it was less prevalent on soils richer in nutrients than on poor soils; check was also most prevalent on clays and least on well-drained sands. On the clays the spruce roots remained almost entirely in the raw humus and did not penetrate the mineral soil, so the trees became particularly vulnerable to drought.

Another factor was also operating. Norway spruce remained healthy in the shade of belts of beech, on moist sites in valley bottoms, and on shady slopes. Thus everything pointed to the exposure of the superficial humus to sun and drought as the basic cause of check. Norway spruce will grow healthily in raw humus so long as the mobilization of nitrogen remains active, but when the humus is dried and heated in the sun it decays only slowly and the supply of nitrogen becomes inadequate. In regions with periods of summer drought where checking of growth has occurred, an admixture of beech would help to keep the stands healthy (Jones 1965).

A careful analysis of growth in first and second generation crops of Norway spruce was made in Denmark by Holmsgaard *et al.* (1961). They examined fifteen paired temporary sample plots in stands aged from 20 to 70 years for the first generation and from 20 to 63 years for the second generation. Most of these stands were on sites that had previously carried beech. The soils were brown forest soils (in three cases) and weakly or well-developed podsols. The soils under the first generation crops had slightly better water-holding capacity and when allowance was made for this, volume production by the two generations was almost identical. The patterns of annual height growth varied markedly with annual variations in climate but without significant differences between generations, Holmsgaard and his colleagues concluded that the case against successive crops of pure Norway spruce was far from proven.

A large amount of research is being done into the processes that link soil and trees in forest ecosystems, into methods of site improvement, and into raising the genetic potential for growth and yield of the trees themselves—all with the object of maintaining and increasing the productivity of sites and stands (see Ford *et al.* 1979; Ballard and Gessel 1983). Cases of reduced yield (defined as the quantity of harvestable biomass accumulated over a given period of time) in successive regular crops of pure conifers and broad-leaved species continue to be investigated, some in great detail as is described on page 18. Despite the very great increase of intensive plantation forestry throughout the world during the present century, examples of decline in productivity remain uncommon.

Foresters accept that they must maintain and improve the productivity of the land in their care. Where silviculture and management based on pure, regular stands are inadequate to protect sites, mixed and irregular stands are used.

The use of mixed stands

Stands composed of mixtures of one or more species may be preferred to pure stands in the following situations:

1. Where mixtures of dead leaves of various kinds produce more favourable humus and so keep the soil in better condition than do pure crops. The evidence for this is strongest for mixtures of broad-leaved species with conifers on soils that are prone to degrade (Troup 1952, p. 195; Jones 1965).

2. Where there is evidence that a mixture will produce a higher yield than the arithmetic mean of the components when grown in pure stands. It is unlikely, however, that the yield of a fast-growing species can be increased by growing it in mixture with a slower-growing species (Johnston 1978).
3. Where the survival and growth of the principal species on soils poor in nutrients is improved by a suitable secondary species (Taylor 1985). Examples are Sitka spruce grown in mixture with Japanese larch on upland sites in Eire (O'Carroll 1978) and oak grown in mixture with Scots pine on upland sites in northern England (Evans 1984, p. 26).

It is believed that the root systems of mixed crops exploit the soil more fully than pure crops. Brown (1986) gives an example for Scots pine and Norway spruce in northern England, but it is not certain that this will always happen. In the case of wind damage, mixing a wind-firm species with one that is shallow-rooting does not necessarily improve the stability of the latter. When considering the incidence of insect attack in forests it is possible, though far from certain, that a more varied and larger population of predators will help to prevent destructive epidemics of insects in mixed crops. It is also believed that mixed crops are less liable to disease than pure crops, but Peace (1961) emphasized the complex effects of mixtures of species on disease organisms and quoted examples of serious damage in mixed crops.

Some forms of silviculture may demand mixtures:

1. Where natural regeneration of the principal species is favoured by the presence of another in the crop, an example being the favourable effect of Silver fir on natural regeneration of Norway spruce (Ammon 1951).
2. In Normandy (France) and the Spessart highlands (West Germany) oak is grown with beech to assist in controlling development of epicormic shoots which cause degrade in the timber of oak (see pages 118 and 188).
3. In areas of outstanding natural beauty where mixtures of evergreen and deciduous species increase visual interest and enhance amenity.

It is often suggested that cultivars produced by tree breeding should be planted in mixtures to offset the risks of disease, insect attack, and damage by other agencies that are believed to exist because of their restricted genetic basis. Zobel and Talbert (1984, p. 271) suggested, from their experience as tree breeders, that the true amount of genetic diversity in forest trees is generally underestimated. All members of a clone of poplar have the same genotype but it may include several genes that confer resistance.

Moreover, parents and offspring are tested for their resistance to disease and insect attack during a tree breeding programme. Despite such precautions, serious damage has occurred to cultivars (and provenances) and the use of several rather than reliance on one or a few provides some insurance against loss. One arrangement, favoured by Zobel and Talbert (1984) and Leakey (1987) is to plant several cultivars or provenances in separate pure stands. If one succumbs it can be replaced, provided that the stands are reasonably large (see Appendix 2). If the mixture is quite intimate, consisting of alternate trees or rows, removing the dead or damaged trees could leave the site poorly stocked.

The use of irregular stands

Regular stands consist of trees of one age class. Regeneration is accomplished over a short period, the canopy being removed in one or a few fellings. In the ideal irregular stand, trees of all age classes grow in intimate mixture. Regeneration is continuous and the canopy is also continuous throughout the stand, both vertically and horizontally (see Fig. 26 on page 164).

The irregular structure may be preferred:

(1) when it is necessary to gain resistance to damage by wind and snow;
(2) where it is necessary to conserve the physical condition and nutrient status of the soil under a permanent forest canopy;
(3) to increase the proportion of timber having special qualities;
(4) to gain certain additional benefits by adopting forms of stand similar to those found in some natural forests.

Stem taper is probably one of the most important single factors affecting susceptibility to breakage by wind (Brünig 1973; Savill 1983). It is also an important factor in snow damage to trees (Petty and Worrell 1981). Trees with low stem taper are more likely to be damaged by wind and snow than trees of similar dimensions with high stem taper. Assmann (1970) compared dominant trees of similar stem diameter from managed irregular and regular stands and found that the former group had markedly greater stem taper. Thus, in areas where damage by wind and snow is to be feared, irregular stands may be preferred to regular stands.

As we have seen on page 8, exposure of some soils by clear felling can lead to deterioration with consequent loss of productivity. Therefore on long steep slopes liable to erosion and on soils that are liable to become degraded it is considered best to preserve a permanent canopy. Those mature, mixed, and irregular forest ecosystems that have been thoroughly studied, such as the Hubbard Brook Experimental Forest in the eastern United States (Bormann *et al.* 1974) are notable for their control over erosion and the close regulation of the nutrients moving in the streams. Hence in the mountainous parts of central Europe and similar regions elsewhere, foresters have worked to create forests composed of stands that are mixed in species, irregular in structure, and provide the protection of a continuous canopy (Fourchy 1954).

The practice in managed irregular forests of retaining trees with the greatest potential and adjusting the proportion of species to take account of local variations in site, may give the flexibility to increase the yield of the growing stock and improve the quality of timber produced. At first the saplings are dominated by the overwood, but this is followed by a gradual relaxation of competition until the dominants emerge with large, full crowns that are well-separated from each other. This pattern of development results in a narrow core of juvenile wood and a large proportion of mature wood with an even rate of radial growth (see page 58). The branches in the upper part of the crown are relatively large but this can be controlled to some extent by careful selection for width of crown in earlier years. The evidence gathered by Assmann (1970) suggested that irregular stands yield about the same or slightly less than regular stands on similar sites, but Köstler (1956, p. 212), Knüchel (1953), and others are consistent in their claims of higher monetary yields due to the superior qualities of the timber.

When forests composed of regular stands are compared with comparable areas of irregular forests, it is evident that certain additional benefits are more readily obtained from the latter (Helliwell 1982). These are a natural appearance, permanent amenity, pleasant surroundings for sport and re-creation, and varied habitat for the conservation of wild plants and animals. Some other advantages that might be anticipated on theoretical grounds from irregular stands are difficult to demonstrate in practice. Murray (1979) and Crooke (1979) both stressed that managerial practices are very important in determining the level of damage by disease and insects in managed forests, but it is not sufficient to assume that mere separation of species in mixed and irregular stands will itself reduce damage (see page 38).

The use of fire in silviculture

Fire is a natural part of many forest ecosystems (Spurr and Barnes 1980) and some depend on fire for their existence. Fire can be destructive under some conditions and beneficial under others (SAF 1984). The effects of fire depend on its intensity; fires of high intensity may kill large trees and damage the top soil, whereas fires that are beneficial are usually of low intensity. Controlled burning is the planned application of fire to natural fuels with the intent to confine it to a predetermined area. Controlled burning is used:

(1)   to reduce the load of fuels, including slash, and so lessen the intensity, spread, and resistance to control of potential wild fires;
(2)   to prepare sites for regeneration by removing some of the litter and humus and exposing the mineral soil;
(3)   to trigger off the release of seed from fruits or cones, reduce a shrub layer and control weeds and unwanted regeneration;
(4)   to reduce infestations of certain insects and the incidence of certain diseases.

Successful controlled burning brings other advantages when establishing a new crop. Fire releases mineral nutrients as ash from the living and dead organisms that are burnt, although some, such as nitrogen, may partially volatilize. Fire converts litter and humus of low pH to ash of higher pH, thus encouraging nitrogen-fixing bacteria more than fungi and stimulating the production of nitrogen.

On the debit side, although fire may be an effective regulator of damaging insects and fungi it also directly affects birds, animals, and beneficial insects and fungi. The practice of concentrating large quantities of slash into windrows or separate piles for burning is likely to increase the intensity of heat and may alter the physical structure of the soil; broadcast burning of scattered slash is therefore normally favoured.

The prescription for a controlled burn should specify the objects of the treatment, suitable conditions of climate and fuel, the frequency of burning during a given season and the probable long-term effects of fire on the soil, the fauna and flora of the site, and the new tree crop. In practice the relative humidity and temperature of the air, wind speed and moisture content of the fuel have most influence on the conduct of controlled burning.

As the experience and skill of managers and workers increases, the range of conditions for safe and effective burning broadens; but the people involved must also be experienced in the control of wild fires and be sufficiently numerous to respond to favourable conditions or contain the fire if conditions do change while the controlled burning is in progress. There is always the risk of people being injured, or even killed, and of equipment being lost (Johnson 1984). In recent years some states of the American union have passed strict fire laws. The clean air acts passed in America and Europe also require forest managers to control smoke which can be a hazard to health.

Controlled burning is widely and successfully practised as a natural and economical means of achieving the purposes described earlier and the techniques used are improving steadily (SAF 1984; Johnson 1984). However, it has also become evident that on certain infertile soils where conservation of organic matter and nutrients (especially nitrogen) is essential for their long-term productivity, controlled burning may have to be ruled out as a means of site preparation.

In southern Australia Radiata pine has been planted on a large scale, notably in the states of South Australia and Victoria, mostly on flat sites with coarse-textured, infertile soils very poorly supplied with nutrients and organic matter, but also on steep slopes with clay loams of moderate fertility. The rainfall is low (around 800 mm) and falls largely in winter. The summers are marked by droughts, high solar radiation, high air temperatures, and low atmospheric humidity. Despite these limitations, first generation crops of Radiata pine have produced a biomass well above that of the indigenous forest.

In the 1960s evidence began to accumulate for a decline in the production of second generation crops, especially on the poor sandy soils. In the first rotation the mean annual increment over a rotation of 35 to 40 years was commonly 20–25 m^3/ha and occasionally reached 30 m^3/ha. Production in the second rotation was, on average, 25 to 30 per cent lower.

Until the 1970s the sites were cleared for replanting by fire of high intensity to reduce the logging residues. Assuming a mean annual increment of 29 m^3/ha, this ground preparation treatment resulted in the loss of 745 kg of nitrogen/ha (Squire and Flinn 1981). The effects of losses of this order on the availability of nutrients to the new crop are not yet fully understood, but the evidence suggested that the method of controlled burning contributed largely to the decline in production.

The principle of retaining litter and logging residues as a source of organic matter and nutrients for the next crop was accepted by forest managers. At Rennick in the State of Victoria, clear felling is completed early in summer so that the residues become dry and brittle and can be macerated by machine during autumn and early winter. The lack of green slash at time of planting is also important to control Black pine beetle (*Hylastes ater*) which can kill large numbers of young plants (see page 41). The macerated residues form a mulch lying over a relatively undisturbed layer of litter. In addition to conserving soil moisture, the mulch also inhibits the growth of natural seedlings of pine which are unwanted because they are very numerous and develop rather poorly.

One-year-old seedlings are planted in mid-winter. A herbicide is applied 6 weeks after planting and any remaining natural regeneration is uprooted

12 months after the planting. Comparison of the growth of first and second generation crops on the same site after 5 years indicated that height growth had not declined (Squire 1983), and this was confirmed by later measurements of volume growth (Farrell *et al.* 1983).

Genetics in silviculture

Since 1937 when C. S. Larsen published 'The Employment of Species, Types and Individuals in Forestry' the principles of genetics have become widely applied in silviculture. The large size and long life cycle of most forest trees have made it difficult to unravel the effects of genotype and environment on the phenotype—the tree in the forest—but since 1950 provenance research and tree breeding have advanced sufficiently to place forest genetics alongside tree physiology and forest ecology as one of the foundations of silviculture.

When natural regeneration is used positive genetic gains can be made, first, by cleaning and thinning the existing crop so as to eliminate undesirable trees and favour desirable ones; and, second, by using some or all of these selected trees as sources of seed, so that they become the parents of the succeeding crop. When techniques of cleaning and thinning such as those advocated by Schädelin (1937) are used to improve the crop (see page 133), the seed bearers will consist of the most vigorous and healthy dominants with straight stems and well-developed crowns, which are well-adapted to the site. This is a form of mass selection. The improvement or genetic gain is the product of the narrow-sense heritability of the individual characters of the seed trees and the intensity of selection, that is, $G=h^2S$.

Heritability is a ratio between genetic and phenotypic variances which indicates the degree to which parents transmit their characters to their offspring. The heritability of wood density in many pines is high, hence the probability is also high that pine trees selected for a given wood density will produce offspring with similar wood density. The figures given for heritability are of two kinds: broad-sense heritability refers to total genetic variance and is exploited when a parent tree is reproduced vegetatively by cuttings or layers; narrow-sense heritability is restricted to the additive genetic variance and is exploited when selecting for seed bearers. The selection differential is greatly influenced by the amount of variance observed in a given character; for example, if there is wide phenotypic variation in stem straightness, from straight through to crooked, the selection differential can be high, but if the majority of trees have straight stems the selection differential can only be low.

In practice the most important improvement occurs because the new crop is well-adapted to the site, adaptation being a highly heritable character (Zobel and Talbert 1984, p. 270). Characters affecting timber quality such as straightness of stem and good natural pruning of branches are moderately heritable, so they can be improved by retaining trees with these characters as seed trees. Heritabilities of characters involved in the rate of growth are low to moderate, so when natural regeneration is practised genetic gains in yield are likely to be small. If higher yields than are given by the existing crop are sought, a superior species, provenance or cultivar must be introduced by planting.

It has been suggested that when natural regeneration is used, the uniform shelterwood or irregular shelterwood systems are superior to the clear cutting and strip systems in the amount of genetic gain that can be obtained. In practice the differences are likely to be small because it is the dominant trees that produce most of the seed. In all these systems, regeneration fellings begin late in the rotation to coincide with the age of maximum seed production, so the forester has ample time to judge the phenotypic qualities of the dominant trees and remove the undesirable ones. Moreover in systems using advance growth these trees will often provide a useful, though incomplete, demonstration of the genetic quality of the dominant trees.

In irregular forests managed under the single tree selection system the measures taken to improve the growing stock ensures that seed for natural regeneration comes from desirable phenotypes. The opposite is the case when exploitation fellings are made to remove the best and leave inferior trees. This dysgenic selection is most severe and damaging when exploitation fellings are repeatedly made to a diameter limit and the smaller and defective trees are left to provide the seed for the new crop (Baidoe 1970; Zobel and Talbert 1984, p. 463). Abundant evidence has been provided by provenance and progeny tests to show that such slowly growing and defective trees are genetically inferior.

The possibility of genetic improvement exists in all silvicultural systems when artificial regeneration is used. When trees are planted they should always be genotypes that best meet the anticipated future uses of the timber. Species, provenances, and cultivars that are superior in rate of growth, stem straightness, branching habit, resistance to disease and insect attack, and tolerance of adverse conditions of site and climate are obtained from seed stands, seed orchards, stool beds and tissue banks. The number of species being improved in breeding programmes has become large and includes many of the principal species used in temperate, sub-tropical and tropical forestry. Economically significant genetic gains are continually being made in many countries (Zobel and Talbert 1984, p. 439; Faulkner 1987; Leakey 1987).

Much of the work on the genecology of forest trees has been done for application in silvicultural practice. Provenance testing has been done for two centuries and it has proved to be a very valuable source of genetic gain. Generally local seed sources of indigenous species are the best adapted to the site and climatic conditions but they are not always the most productive. Introduced provenances may grow faster and have better stem form or other useful characters, but they are usually less well-adapted. Continual natural selection takes place at the new site, which can be aided by thinnings that remove less well-adapted trees. If many provenances are introduced the pool of genetic variance is augmented, and selection followed by controlled cross-pollination within and between provenances that are best adapted and have other useful characters can be used to produce new cultivars.

Since the 1930s a striking feature of provenance research has been international co-operation to establish provenance tests, in which the natural range of important species such as Norway spruce and Douglas fir has been thoroughly sampled and seed of the different provenances has been distributed to many countries. Information about the performance relative to standard provenances grown by all or groups of the participants is circulated regularly. The International Union of Forestry Research Organizations has encouraged this co-operative provenance research which now includes many species important in tropical forestry (Barnes and Gibson 1984).

Unless precautions are taken there is some risk that improvements made through the use of provenances and cultivars of seed and clonal origin may be lost through damage because genetic diversity is reduced. This is a special case of pure, regular stands which was discussed on page 14.

3

The protective functions of managed forests

As already noted, the functions of managed forests may be classified as productive, protective, and social. The most important productive functions are to sustain supplies of timber and minor products, such as foliage, fruits, honey, dyes, and medicinal plants. The protective functions are concerned with the environment and take several forms, including:

(1) protecting the physical features of the land;
(2) regulating the quantity and quality of water supplies;
(3) conserving wild plants and animals through management of their habitats; and
(4) protecting landscapes of outstanding natural beauty.

Forests also provide shelter for crops, livestock, and buildings and this aspect has led to the development of agro-forestry systems (see Chapter 20). We may also note in passing that the principal social functions of managed forests are creating employment and providing facilities for field sports and outdoor recreation. These services flow from the management of forests for production and protection and we return to them on page 60.

The relative importance and compatibility of the productive and protective functions depend on the conditions of climate, terrain, and soil that prevail in each area. Over large expanses of land there is little or no conflict; timber production has priority and the protective functions are associated with it. The growing stock generally consists of regular stands, often pure but sometimes mixed. When the site conditions are such as to make protection essential, as for instance in forests on important water catchments, timber production may become an associated function. Such forests consist of regular or irregular stands and may also be composed of certain acceptable species. Where site conditions are very severe the protective functions have priority; timber production is subordinate to them and may even be ruled out altogether; these are the true protection forests. Within the boundaries of managed forests there may also be geological features, archaeological sites, and small areas of natural woodland which are all left undisturbed.

Protection forests in mountainous regions

The silvicultural systems used in mountainous areas have evolved slowly over several centuries. They usually have four objects:

(1) to provide protection against avalanches, landslips, erosion and floods;
(2) to protect the habitat of native fauna and flora;
(3) to safeguard the high quality of the landscape; and
(4) to ensure the livelihood of the people, as a consequence of the first three objects.

In mountainous regions throughout the world there are several physical hazards and difficulties to be faced. The steep slopes and scarps hinder access for silvicultural operations and make extraction difficult; the climate also shortens the period available for these operations. Individual stands are menaced by snow, strong winds, rock falls and landslips. Restocking is made difficult by infrequent seeding and competing vegetation, and young trees may be damaged by creeping snow. Favourable climatic cycles are needed to induce seed production and make restocking easier. The sites are very variable. They occur as a mosaic varying in terrain, exposure to wind, and soil depth over quite small areas. The stands are the units of silviculture and management and each unit is as homogeneous as possible, the size ranging from 5 to 20 ha (Dubourdieu 1986a).

The factors affecting choice of silvicultural treatment are, on the one hand, the condition of the stand—species composition, structure, age and probable future longevity, and, on the other hand, site conditions and the need to safeguard the physical environment and landscape. If a stand tends to an irregular structure and is expected to survive beyond the planning period, irregularity is favoured, particularly where the need for site protection is great. If the site is less severe, or the life expectancy of the trees is short, a correspondingly short regeneration period is prescribed and the stand structure becomes regular. The guiding principles are flexibility of treatment and strengthening the existing growing stock.

Forestry on water catchments

The land on which forests are growing is often also used as a source of water, particularly when the supply of water flowing from the forested catchment is considered to be reliable and of high quality. Forest managers must always strive to conserve such supplies and three aspects of the influence of forests on water supplies requiring attention here are total yield of water, regularity of supply, and water quality.

We have seen (page 8) that water is lost from a forest canopy partly through evaporation of intercepted water and partly by transpiration. Because of the turbulence imposed on air flowing over forests by the roughness of the upper surface of the canopy, evaporation is higher from forests than from vegetation with smoother canopies (Miller 1984), and the yield of water from a forested catchment is reduced. The amount of the reduction depends on the pattern of rainfall and air temperature and also on the proportion of the catchment that is occupied by fully stocked forest. When stands are felled the yield of water rises, and when felled areas are restocked the yield gradually falls. The actual responses of a catchment to forest operations depend on its geomorphology and hydrological characters and in consequence are difficult to predict in detail.

Nutter (1979) emphasized that most of the information about peak flows was derived from catchments bearing natural forests, but he considered that those bearing managed forests were unlikely to behave very differently. The information is that peak flows are lower and later from forested catchments than from grassland. Nutter went on to suggest that one way to reduce the impact of forest operations on water yield is to disperse thinnings, fellings, drainage cultivation and restocking over a catchment, so that increases in water yield in one part are compensated by reductions in another part. This aspect of management is taken up again on page 50.

Turning to water quality, supplies from natural forests are low in suspended solids, low in temperature, high in dissolved oxygen and low in nutrients, pesticides, and other substances leached from the soil. The water quality from different catchments does vary somewhat because it reflects climate, species composition of forest stands and other vegetation, the underlying geology, and the soil type. Forest management decisions are superimposed on these factors.

Erosion is responsible for the solid particles in streams, and the principal sources of erosion are the construction of forest roads, harvesting operations, opening new drains and cleaning old ones, and soil cultivation. Rowan (1977) described a classification of terrain using as key features the bearing capacity of the soil, the presence of obstacles to movement across the surface of the land, and the gradient and topographic form of the slope. The classification is recorded on maps which are used to plan road networks and methods of harvesting. When a felling is to be done, suitable harvesting machines are selected and working procedures devised to prevent undue disturbance of the site.

Forest ecosystems are particularly effective at intercepting and retaining natural inputs of nutrients from the atmosphere (Miller 1984). Nitrogen, phosphorous, and potassium applied in fertilizers to mineral soils are usually efficiently retained. Thus water draining from forests removes very small quantities of nutrients, other than calcium and magnesium derived from soil weathering. The retention of nutrients by forest ecosystems depends largely on uptake by the trees, so when a stand is clear felled there may be contamination of streams unless the site is quickly colonized by ground vegetation, and restocking with trees is also rapid. In these circumstances the peak of soluble nitrogen is low and of very short duration.

The actions taken to protect water supplies from forested catchments may include the following:

1. Road networks are designed and constructed to standards that minimize erosion.
2. Extraction of timber is done by methods that cause minimum possible soil disturbance.
3. The water courses are protected by buffer zones of natural vegetation at least 15–20 and up to 30 m wide. Water from drains and plough furrows is not discharged directly into the streams but first passes slowly through mineral soil to remove silt and nutrients and neutralize chemical contaminants.

Some other actions taken to protect water supplies have been useful under certain conditions but may not be suitable for wider use. Examples are the substitution of broad-leaved species for conifers, the use of heavy thinnings to reduce the stocking of stands on catchments, and retention of narrow strips of mature trees alongside streams. Controlled burning of slash has minimal effects on water quality unless the ground vegetation and litter layer are destroyed and water reaches the streams directly over the soil surface (see page 16).

Conservation of wild plants and animals

The purpose of nature conservation is to ensure survival of native species in their habitats (Boyd 1987). In a given region, a forest managed primarily for the sustained yield of timber from one or a few species usually provides a smaller range of habitat for wildlife than the natural indigenous forest. Nevertheless, managed forests do provide habitat for a significant part of the native flora and fauna.

Many wild species can adjust to and even benefit from disturbances caused by thinnings, regeneration fellings, and clear fellings (Rochelle and

Bunnell 1979). Species that disperse and colonize rather slowly, or have small individual populations, or whose numbers are subject to large fluctuations are more likely to be lost and may need protection (Peterken 1977). Certain silvicultural and managerial practices can be used to improve and diversify the habitat for wildlife in general and protect the less mobile or less adaptable native species.

In a managed forest the occurrence and activity of wild plants and animals is affected by:

(1) the extent and condition of zones of transition between the external edges of closed forest and other natural or man-made ecosystems bordering the forest;

(2) the occurrence of clearings of various kinds which form internal edges within the forest;

(3) changes in ground vegetation, stand structure, and microclimate within stands during the rotation; and

(4) management decisions affecting length of rotation and forest operations.

The external boundaries of a forest may march with alpine meadow, open moorland, arable or pastoral farming, fresh water, or even an ocean. The interactions between the forest and another ecosystem in the zone of transition between them can create conditions more favourable to wild plants and animals than the sum of their individual contributions (Boyd 1987). Whenever possible the external edges of a forest should not be abrupt or sharply defined. Opportunities for creating transition zones are often greatest when different land uses are practised under one ownership, as on large, privately-owned properties.

Internal edges generally lie within the forest manager's control. They separate productive stands and open spaces of various kinds. Forest roads create major internal edges, occurring as ribbons of disturbed ground with good drainage. One common practice is to halt productive crops at distances of 20 to 80 m from a road to provide space for ground vegetation, shrubs, and trees in a mixed and irregular edge providing habitat for plants, insects, reptiles, and birds.

The secondary parts of the road system, called rides, are not always wide enough to produce a good range of habitat. Indenting or scalloping the edge of productive stands to form opposed or alternating bays, and forming glades at intersections with other rides or roads will provide well-lit sites for wild plants and butterflies. Retaining old trees as standards and maintaining small ponds near the rides will further diversify the habitat.

Open water within a forest becomes a focus for animals and people especially if a generous transition zone is developed. The range of plant communities from those of deep water to marshy and drier ground may be complemented by native shrubs and small trees before productive stands are reached. Such varied habitats provide feeding, breeding, and over-wintering sites for many birds and animals. Similarly the courses and banks of rivers and streams can be made rich in riparian habitats and interesting species of plants. A zone of at least 15–20 m is often treated so as to provide permanent reserves of vegetation which protect the banks and create favourable conditions for fish. The streams themselves may also be im-proved using methods such as those described by Mills (1980).

The value of a given stand as habitat for wild plants, animals, and birds is greatest at the beginning and end of the rotation. During the regeneration stage slash provides local shelter; cultivation and drainage may bring minerals and grit to the soil surface. Forage is abundant so game birds, small rodents, and large mammals all benefit. Insectivorous birds increase and predators appear. When the tree crop enters the thicket stage, much of the ground vegetation is suppressed and mice and voles decline. However, in the temperate regions foxes (*Vulpes vulpes*), badgers (*Meles meles*), and deer take advantage of the warm cover. During the pole stage, thinnings increase light and ventilation so ground vegetation reappears and forage becomes more plentiful. Insects increase in species and numbers as do insectivorous birds. Mature stands provide lairs, breeding sites, roosts, and perches for birds of prey and food for seed-eating birds. The number of species of song birds is greater in stands with several canopy layers than in those with only one layer, and in general mixed and irregular stands attract more species of birds than pure, regular stands.

Where the principal function of a forest is to produce timber of one or a few species, improvement of habitat for native wildlife is an associated function. Cadman (1965) gave examples of what can readily be done in production forests. Small communities of rare plants, which might be overcome by vigorous regrowth of vegetation during regeneration opera-tions, may be marked for protection from disturbance. A badger's sett may be left unplanted. Traditional rutting stands of Fallow deer (*Dama dama*) are also left and the main tracks of Red deer (*Cervus elaphus*) incorporated into the network of rides and glades for observation, control, and winter feeding.

In a forest managed for sustained yield of timber, adjacent stands differ in age and height, species composition, and canopy development. Particu-larly favourable edges occur where a stand nearing maturity adjoins an area being regenerated—conditions that are created continually by the clear cutting and strip systems of silviculture (see pages 65 and 138). When the growing stock of only one species approaches the condition required for sustained yield, the land is covered by a mosaic of stands differing in area, age, height, and canopy development. This mosaic is linked by roads and rides and interspersed by clearings of various kinds, many being capable of providing habitat for native plants and animals. If the forest is situated in hilly country there will often be farmland below and open land above; the ridges and valleys may also carry native flora and fauna.

When greater importance is given to the conservation of wildlife a conservation plan is compiled. An inventory is made of the important habitats, such as fresh water, moorland, grassland, wetland, or indigenous woodland and of the rare or otherwise interesting species occurring in them. This inventory may be compiled in association with conservation specialists. Where the species and their habitats are particularly valuable, arrangements are made to control access or reduce disturbance at critical times. The range of habitat in the forest is increased by retaining some stands beyond the normal rotation age, encouraging development of understorey species and an irregular structure, and establishing mixed stands containing species that produce edible fruits and seeds.

Where natural communities of plant and animal species occurring in the forest are of national or international importance and must be protected from loss (as is the case in many areas of tropical moist forest), parts of the forest may have to be set aside as nature reserves. Normal felling and harvesting ceases and the community is left to develop naturally. Ideally the area of the reserve is made sufficient to permit establishment of a protective buffer zone around a central undisturbed core. An example is provided by Rannoch forest in the central highlands of Scotland. Here the greater part is given over to timber production, but there is also a reserve of indigenous Scots pine woodland which is surrounded by a protective zone (Godwin and Boyd 1976).

Silvicultural systems in the landscape

Most people like trees, perhaps because part of the evolutionary development of many races took place among trees and forests (Davis 1976, p. 260). In most European countries active concern about trees is widespread, and the peoples of some Asian countries worship trees for their size and beauty. Misunderstandings about the methods and purpose of organized forestry can occur unless steps are taken to forestall them. Because trees are long-lived and the span of human life is shorter, people tend to think of them as permanent and of forests as unchanging. In modern industrial societies there is a general lack of public recognition that forests and trees can be managed for sustained yields or that forests are sources of timber, paper, chemicals, and many other products used in their daily lives. Moreover, as the population of a country becomes concentrated in urban areas and is more affluent, the pressures caused by their restricted living space stimulates demand for outdoor recreation and increases interest in wild plants and animals. All these factors can lead to controversy about forestry which must be considered when applying silvicultural systems and in forest planning generally.

To the general public a clear cutting looks like forest destruction rather than as a prelude to forest renewal. In hilly or mountainous areas clear cuts and strip fellings can be highly visible to the public and if such felling coupes are laid out without sensitivity they can become an affront, especially when slash is plentiful and regeneration takes several years to show. In contrast, forests worked under one of the irregular systems (irregular shelterwood, single tree selection, and group selection) can have a natural and unchanging appearance, which makes them suitable where trees form an essential part of a well-loved landscape.

The visual impact of several silvicultural systems can be made more acceptable by considering the size, shape, and arrangement of coupes in relation to the visual character of the landscape. According to Crowe (1978), the configuration of the terrain and the scale of its variation, the existing type and pattern of vegetation and land use, and the prevailing colours of rock, soil, and structures all influence the visual character of the landscape. The first step is to analyse the general type of regional landscape. Regions with large-scale rolling hills can accommodate large coupes; whereas in regions with small-scale contrasts of land-form and delicately modelled hills, each coupe must be treated individually and fitted into the intricate pattern.

Now follows more detailed inspection of the site from different view-points to determine the shape of coupes or compartments. The upper margins should be related to topography so that it accentuates the form of the land. Features such as the dip of the underlying strata, outcrops, cliffs, streams, waterfalls, and gullies should also be respected when fixing the upper and side margins of coupes or compartments. If the forest adjoins agricultural land the lower margin may follow the same contour. Often a fairly complex shape of coupe with inlets and promontories will fit into the landscape more easily than a large, simple shape (Fig. 6, page 60).

The complexity of the environmental impact of silvicultural systems has led to the practice of designing forests so that the needs of regeneration, protection of the growing stock, efficient harvesting, conservation of wild-life, protection of water supplies, and enhancement of scenic beauty are all considered together (TGUK 1985). We must now turn to forest protection before considering forest design on page 59.

4

Protecting forests against damage

Four kinds of damage are considered here. These are:
1. Damage caused by extremes of climate, particularly wind, snow, and fire.
2. Damage caused by certain diseases, insects, and animals.
3. Damage to site and trees in the course of forest operations.
4. Air pollution

These kinds of damage are recurrent and can disrupt the orderly regeneration, tending, and harvesting of tree crops unless control measures are devised and applied. Injuries to site and trees reduce the productive potential of the forest by depressing tree growth and lowering the volume and value of timber produced. When the cost of protective measures is combined with loss of growing stock and increment these climatic, biotic, mechanical, and environmental agents can place a heavy financial burden on a forest enterprise. The purpose of this chapter is to show how the risk of damage may be assessed and how the resistance of the forest can be raised so that the impact of these agents on silvicultural systems is kept as low as possible.

Wind

In many parts of the world, especially in those regions with maritime climates, wind damage is a recurring risk. Strong winds can uproot trees, break their stems, and cause stems and crowns to become misshapen. Such winds reduce growth increment and make difficult the regeneration, tending, and harvesting of tree crops. The steps required to reduce damage by strong winds are, first, assess the risk of wind damage; second, study the effects of wind on tree crops; and third, adopt measures to increase the resistance of a forest to damage.

Assessing the risk of wind damage

Two categories of damage are recognized, called catastrophic and endemic (Booth 1977; Miller 1985). Catastrophic damage is caused by winds of unusual intensity such as blew in January 1953, 1976, and 1968 in Britain, in November 1972 in western Europe, and in October 1987 in south-east England and western Europe. The damage caused by these very strong gales was influenced more by the extremely high wind speed, the direction of the wind, and local topographic features than by soil conditions (Holtam 1971). It is not possible to predict when and where future catastrophic wind damage will occur, and the sole means of moderating significantly the effects on a forest is to increase the diversity of the stands in age and height, so reducing the proportion of the forest area that is at risk. This is an important argument for creating normal forests (see page 52). A similar solution is also appropriate in several other parts of the world. In China, the Philippines, and Japan typhoons cause catastrophic damage to forests; in the Caribbean, Gulf of Mexico, and south-east coast of North America the damage is done by hurricanes; and in parts of the Indian ocean tropical cyclones produce the very heavy rainfall and intense winds which may gust up to 60 m/s (120 knots).

Endemic wind damage is caused by gales with a relatively moderate wind speed of 20 m/s, gusting to 30 m/s. In the British Isles and north-western and central Europe such gales occur several times each year, mostly during the winter. A common result is the occurrence of fresh windthrow in the less stable parts of crops which are approaching maturity. The damage is strongly influenced by site conditions and silvicultural practice and it is possible to reduce significantly its incidence and extent. Where damage by gales is to be feared the risk can be assessed by examining:

(1) the windiness of the climate of the region;
(2) the elevation of the site;
(3) the size, shape, and alignment of features of the terrain; and
(4) the condition of the soil in relation to the development of tree roots.

From long-term meteorological records it is well known that the north and west of the British Isles, western Norway, Jutland in Denmark, and northern Germany experience strong winds more frequently and at greater strength than other parts of Europe. In addition coastal areas have more frequent gales than inland areas. Booth (1977), Miller (1985), and Miller *et al.* (1987), divided Britain into zones by the incidence of strong winds. Where extensive forests lie close to or straddle the boundaries of these wind zones, local surveys of exposure can be made by setting out 5 to 10 flags of standard dimensions and texture and measuring their rate of tatter at intervals of 2 months for a period of 3 years (Miller *et al.* 1987).

Mean wind speeds increase with elevation, as does the frequency of gales; thus, forests at higher elevations are generally more prone to wind damage than those on lowland sites within the same geographic region. In Britain, rainfall also increases at higher elevations so that soils are wetter and their penetration by tree roots is often reduced. These effects become progressively stronger so that, for a given increase of elevation, there is a more rapid rise in the incidence of wind damage at higher than at lower elevations.

The exposure of a site to wind is modified by the size, shape, and alignment of adjacent features of the terrain. The relative exposure of a site can be assessed by measuring the angle of inclination to the visible horizon at the major compass points; the sum of the eight measurements of angle, obtained by clinometer, is called the topex value (Malcolm and Studholme 1972; Pyatt 1970). A site on a flat coastal plain usually has a topex value of 0, confirming that it will be severely exposed. A site completely surrounded by higher ground will have a topex value of 60 or more, indicating that it is moderately or very sheltered. If a site lies in a valley open to and aligned with the direction of the prevailing wind it will be relatively more exposed than if the axis of the valley lies at right angles to the wind. Sites on long, moderate slopes are often exposed, usually regardless of their aspect, and the higher sites will often be severely exposed because the wind becomes stronger and more turbulent as it blows up the slope. Mountain ridges, isolated rounded hills, and shoulders are all liable to be severely exposed (Hütte 1968). Most damage is done to forests on the windward side of topographical features but in mountainous areas damage can also occur on the leeside; this has been reported from Europe by Hütte (1968) and New Zealand by Hill (1979).

For areas with varied terrain it is necessary to survey a full range of elevations and aspects, using a systematic sample, to determine local changes in the direction of the prevailing wind. Assessment of topex values can often be combined with soil surveys which also assist in assessing the likelihood of wind damage. Booth (1974) placed models of the terrain of some upland forests in wind tunnels, so as to learn more about the flow of wind over and around topographical features and identify sites with high risk of wind damage.

The depth, extent, and morphology of the root system are influenced by the volume of rootable soil and its physical condition, particularly moisture content and aeration (Sutton 1969). Where rooting is unrestricted to a depth of 45 cm, as in brown earths and podsols, the risk of windthrow is low (but there is risk of stem breakage during severe gales). If rooting is restricted but there is some penetration of structural roots to depths greater than 25 cm, as is common on deep peats or loamy gleys, the risk of windthrow is greater. Where rooting is restricted to a depth less than 25 cm, as is usual on peaty gleys, shallow indurated soils, and waterlogged soils the risk of windthrow is high or very high.

The assessments of regional windiness, elevation, topex, and soil are combined to give six windthrow hazard classes which are displayed on maps and used in planning the application of measures to reduce damage.

The effects of wind on tree crops

The onset of wind damage depends on the interaction between turbulent wind passing over the forest and the dynamic response of the stands, comprising trees of varying sizes and shapes which are imperfectly anchored by their roots to the soil. When a tree is bent over during a gale, the total bending moment applied about the base of the stem arises from the frictional drag of the crown and the weights of the stem and crown. Because wind speed decreases rapidly from the top of the canopy downwards, the greater part of the applied moment due to wind drag is contributed by the upper half of the crown (Petty and Swain 1985). The elasticity of the stem wood resists the total applied bending moment so that when wind speed is reduced the bent tree springs upright. During a gale the trees sway and although the swaying is partially damped by mutual contact between the crowns of neighbouring trees, swaying over long periods of time may cause the soil beneath the stems to break; it may also gradually loosen the roots in the soil. The resistance of the root system to the total applied bending moment appears to depend mainly on the strength of the soil in tension, the strength of the roots on the windward side of the tree as they are placed in tension, and the weight of the anchoring plate of roots and soil (Coutts 1986). When the total applied bending moment exceeds the maximum resistive moment the stem will break, but on gleyed soils it is more common for the soil and roots to break, so the tree is uprooted.

Stands become increasingly susceptible to wind damage as they grow taller. In Britain the top height of a stand when 3 per cent of the live trees have been thrown by wind is termed the 'critical height', which can range for unthinned stands from 13 m on sites with the highest windthrow hazard to 28 m where the hazard is lowest (Miller 1985). At first, scattered trees or small groups are blown down. The damage then spreads progressively during several years. When 40 per cent of the crop has been lost, 'terminal height' has been reached and the remaining trees are clear felled.

As already noted on page 15, stem taper is a very important factor affecting susceptibility to breakage by wind and snow. If stem taper is expressed as the ratio of tree height to stem diameter at breast height, Sitka spruce trees 20 m tall with taper 1:100 would certainly be unstable, whereas trees of taper 1:60 should be very stable. The ratio of crown weight to stem weight is also important—to minimize damage a small value is desirable. So, provided this ratio does not vary greatly with stem volume, trees of higher stem taper grown at wide spacing should have greater stability than trees of lower taper grown at narrower spacing. The former is the condition of dominant trees in the upper canopy of managed irregular stands and, as we have seen on page 15, they are resistant to wind and snow. Conversely, when a regular stand is kept densely stocked during the first half of its life, the live crowns become short and the stems have low taper. If such a stand is thinned as it nears the critical height on an exposed site, increased penetration of the upper canopy by strong winds will do much damage.

Soil-borne fungi which attack tree roots causing their decay, reduce the anchorage of infected trees to the soil and increase the risk of windthrow. The most common are species of *Armillaria*, *Phaeolus schweinitzii*, and *Heterobasidion annosum* (see page 40). *Polyporus dryadeus* can cause death of structural roots of oak (Peace 1962), and in Britain attack by *Rhizina undulata* kills the roots of conifers, especially spruces (Murray and Young 1961). The trees then die in groups and the small gaps created in the canopy become the source of more extensive wind damage.

Increasing the resistance of forests to wind damage

There are several possible approaches to reducing endemic wind damage and increasing the resistance of the growing stock to strong winds. Most of the techniques used at present are the result of long periods of trial and error. Much remains to be learned about the behaviour of wind as it flows over forests and about the reaction of various tree species to it. The mathematical and physical approaches now being used in research give promise of further advances in silvicultural techniques (see Petty and Swain 1985; Coutts 1986; Thomasius *et al.* 1986; Blackburn *et al.* 1988).

Wind damage to forests commonly originates at recently felled margins facing the wind; in small clearings created for regeneration or caused by death of trees; at edges associated with changes in top height of 5–8 m or more (Somerville 1980); in poorly drained areas; and where the alignment of roads or rides alters direction sharply. Most of the counter measures are long-term so as many as possible are incorporated into the basic design of the forest (see page 59).

The silvicultural practices that are common in windy climates include the following:

1. Fellings that suddenly expose the interior of a stand to the prevailing wind are avoided. The tree crops are arranged so that fellings in

successive adjacent areas proceed against the prevailing wind direction. This ensures that there will always be a sheltering stand or good shelterbelt to windward of the felled coupe.

2. Large clearings and small scattered coupes within mature crops are also avoided. When narrow coupes, similar to cleared strips, are used they are normally aligned at right angles to the direction of gales.

3. Thinnings are confined to the early stages of stand development. They begin during the thicket stage or even earlier and end before the crop reaches the critical height for the site.

4. Seed trees to be used for regeneration are identified as early as possible and are gradually released from competition by their neighbours so that their crowns develop and their stems taper strongly (Smith 1986).

The permanent edges associated with the external boundaries of the forest and with roads, rides, cutting sections, and fire breaks within it are specially treated, because wind turbulence generated by the edge trees can cause damage to stands to leeward of them. It is usual to treat permanent edges in a similar manner to shelterbelts so as to create stable margins 30–50 m wide which the wind can penetrate, so reducing turbulence behind them. They are thinned heavily in early life, and after attaining heights of 15 m are left unthinned. Deciduous species may be used for these permanent margins and high pruning of evergreen species is sometimes done to increase permeability to wind (Savill 1983). The treatment of edges to encourage wildlife and for fire protection is described on pages 25 and 37.

In hilly country the prevailing winds tend to become diverted according to the trend of the valleys and the local wind direction must be identified. In central Europe narrow coupes arranged at right angles to the local wind direction have been used for centuries. One of the most dangerous local winds sweeps over a plateau from the west and drops down the eastern slope flanking it. On slopes swept by a descending wind of this kind it is customary to adopt vertical coupes. Again, on the southern slopes of certain mountains in Austria, the arrangement of coupes must also guard against the west wind as well as a wind descending from the mountain tops (Troup 1952, p. 83). The solution requires fellings on the lower sections to proceed in advance of those in the sections above them, so that there is always a sheltering stand on the uphill side of the exposed vertical coupes. The use of felling keys as a guide to the alignment of strips and the direction of felling in hilly country is described on page 141.

Where the windthrow hazard is high it is essential to stimulate development of root systems, which are as deep as possible and well-distributed radially. There are many sites where cultivation and drainage will improve the moisture content, aeration, temperature regime, and strength of the soil, thus encouraging the growth of roots and increasing the resistance of the trees to windthrow. The value of these site preparation treatments for species like Sitka spruce is further enhanced because the basic structure of the root system is established early in life (Coutts and Philipson 1987). Drainage is likely to be most effective in producing conditions for better root development on surface-water gleys of sandy or loamy texture, but the clay-gleys with very low hydraulic conductivity are difficult to improve by draining or subsoiling alone.

Adoption of measures to control the spread of fungi causing death of roots is another approach to reducing damage by wind. The extension of group-dying associated with *Rhizina undulata* can be avoided by not lighting fires in or near developing stands, maintaining drains, and not thinning near infected sites. In Britain, it is possible that controlled burning to prepare sites for restocking will encourage attack by *Rhizina undulata* (Gregory and Redfern 1987).

Snow

In the boreal forests of Scandinavia and North America and in the mountain forests of central Europe and northern India snow lies for several months of the year. It can protect young trees during winter but on slopes the slippage of compacted snow can cause bending and breakage of stems. The stems and crowns of trees in the thicket and early pole stage are liable to breakage by snow, especially when cleanings and thinnings are neglected or delayed and stocking is dense. In Europe larches, pines and spruces, and in north-west America and Canada Western hemlock and Ponderosa pine are prone to stem bending and breakage. The economic loss can be severe.

When certain south coastal provenances of Lodgepole pine were planted on upland sites in north-east Scotland their lax habit and poorly developed root systems made them very vulnerable to damage by snow. Petty and Worrell (1981) and Thomasius *et al.* (1986) have shown that trees with straight and strongly tapering stems and long live crowns are more resistant to snow break than leaning trees and those with short crowns or crooked stems. Selecting and favouring well-shaped dominants is essential when the risk of snow break is high and as we have already noted on page 16, one important justification for irregular stands in the mountain forests of central Europe is their resistance to damage by snow.

Fire

The danger of wild fire arises from a combination of fire risk and fire hazard (Hibberd 1986). The risk or likelihood of a fire starting increases with the presence of people in the forest, the occurrence of fires on adjoining land, and where roads and railways pass through a forest. Fire hazard refers to the quantities of vegetation and slash in the forest and the flammability of this fuel. Fire hazard is high when the relative humidity of the air is low, air temperature and wind speed are high, and the moisture content of the fuel is low. The danger of wild fire becomes extreme when high risk is combined with high hazard.

The first step in fire prevention is creating an organization which ensures speedy reporting of outbreaks. The next is provision of effective means for suppressing fires, and the third step is prevention and control of fire. It is the silvicultural aspects of prevention and control which concerns us here.

The dominating influence in fire hazard is climate. Wild fires are a recurrent danger in regions with long periods of hot, dry weather, such as the Mediterranean basin, the African savannas, southern Australia, and southern California. A typical sequence in the development of wild fire in the Mediterranean region is described by Delabraze (1986). It begins with a ground fire, often in neglected scrub or derelict coppice on the edge of a forest. If there are small- and medium-sized trees with flammable foliage to windward of the source the ground fire soon develops into a crown fire and, as the conflagration intensifies, flaming brands are released and are borne by the wind to spread the fire. Belts of tall trees of relatively fire-resistant species can intercept these brands and provide fronts from which to fight the fire.

Planning to reduce the danger of fire and increase the resistance of a forest to damage begins with a study of climatic patterns and the phenology of vegetation, in order to identify periods of greatest danger. Analysis of the terrain and assessment of exposure to the prevailing winds in the most vulnerable parts of the forest permits identification of points where a fire is likely to accelerate or slow down (see page 32). The vulnerability of each stand to fire damage is then assessed. Wherever possible existing tall stands of fire-resistant species are incorporated into fire belts or barriers. Regular treatments (including soil cultivation, application of herbicides, mechanical mowing of vegetation, grazing by animals, and controlled burning, as appropriate) are applied to reduce the amounts of fuel on the external boundaries of the forest and along the main extraction routes; the latter can then become internal fire breaks. Areas of inflammable scrub and ground vegetation are cleared at points of high risk and a network of clear fire breaks is established to prevent fire extending unimpeded through the forest. Fire breaks perpendicular to the prevailing wind are sited to leeward of hills and tall stands; those parallel to the wind lie along crests and are often made zig-zagged so the wind is not channelled along them. Where possible the fire breaks should be close to water courses, ditches, walls, and similar obstacles. During the fire season the main extraction routes and fire breaks are kept unobstructed so that fire suppression teams and their appliances can reach fires quickly and can also retreat to more secure places.

It is common experience that the progress of a fire can be checked by a stand of another species or age class, so silviculture aims at producing mosaics of small stands, often of several different species (Delabraze 1986). Those parts of stands that face the winds prevailing during periods when risk is high are kept densely stocked so that any understorey and ground vegetation is suppressed. Those parts facing away from the wind can be thinned more heavily. Felling coupes are arranged so that a ground fire will not readily be propagated into a crown fire by a steady succession of taller stands downwind; instead the sequence of age or size classes is as varied and distinct as possible (see pages 60 and 83).

Fire is one of the greatest dangers to organized forestry in the savanna regions of Africa where grasses, leaf and needle litter, and slash provide much of the fuel (Laurie 1974). Controlling growth of grasses within stands requires thorough site preparation to eradicate them before planting, intensive weeding until the canopy closes and light thinnings thereafter, so that regrowth of grasses is suppressed. However, soil water often becomes limiting making heavy thinnings essential, so grasses reinvade and the fire hazard rises. If the tree species is resistant to fire the amount of slash left by pruning and thinning can be reduced by controlled burning late in the wet season or early in the dry season, and after the hottest part of the day. Controlled burning sometimes begins after the first pruning when the trees are 8–11 m tall. In stands nearing maturity, successive controlled burns are done until the quantity of fuel falls below 12.5 t/ha (Laurie 1974). The maceration of large quantities of slash was discussed on page 18.

Disease

Throughout the rotation in managed forests the incidence and activity of fungal pathogens is affected by four factors (Murray 1979). These are:

(1)   the past history of the crop and the residues from the previous crop;
(2)   pathogens carried on planting stock raised in nurseries;

(3) changes in the microclimate within stands during the rotation; and
(4) management decisions and forest operations.

Because the incidence and impact of damaging fungi are often associated with a particular stage of the rotation, it is convenient to present this section in the four stages of: establishment, thicket, pole stage, and maturity.

Pathogens of the establishment stage

If regeneration is by planting, many of the pathogens of the establishment stage and early life of the crop may originate in the nursery and be taken into the forest by infected plants. Examples are provided by the blister rust *Cronartium ribicola* of five-needled pines, the more pathogenic species of *Lophodermium* on Scots pine and *Scleroderris lagerbergii* on Red and Jack pines in northern USA and Canada. Some common foliar diseases of the nursery, such as *Meria laricis* on larch and *Didymascella thujina* on Western red cedar become especially damaging on young trees recovering from the shock of transplanting into the forest. Thus, control of disease in the forest begins with disease control in the nursery.

When natural regeneration or direct seeding are used, the young crops are affected mainly by pathogens already present in the forest, either on the site being regenerated or coming in from other parts. Similarly, when nursery stock is planted the young trees are also at risk from pathogens in the forest. Older lightly infected trees may be the source of inoculum which can produce serious and sometimes lethal infection of young plants. Examples are the Brown spot fungus *Scirrhia acicola* in pine stands of the south and west United States and *Nectria ditissima* on beech in Europe. A frequent source of damage originating from the establishment stage lies in the residues of previous crops, particularly stumps and roots which can carry pathogenic root fungi. *Heterobasidion annosum* is a very important example and the means adopted for its control are described later.

Pathogens of the thicket stage

In the thicket stage, the microclimate favours spore casting and germination, which may lead to primary infection of trees that are predisposed to it. As we have already seen (page 7), the developing canopy intercepts increasing proportions of precipitation and solar radiation; wind speeds are also reduced within thickets and the humidity rises to high levels for long periods. The intense competition for water between individual trees in a thicket favours pathogens that can attack stressed trees. Consequently diseases of foliage and bark are common at this stage.

Pathogens of the pole stage

When the pole stage is reached, the major tending operations of brashing, pruning, and thinning begin and wounding of stems and roots also begins (see page 46). Competition between trees is relieved, and the canopy becomes more open and ventilated so the microclimate is less favourable for pathogens requiring moist conditions on leaves or needles for germination and infection. Things are different below ground. The mat of roots has become almost continuous and direct contacts between roots can favour the spread of diseases such as Honey fungus *Armillaria mellea*. When root grafting occurs, wilt fungi such as the Oak wilt *Ceratocystis fagacaerum* can spread.

Heterobasidion annosum is a white rot fungus causing butt rot, root rot and on certain sites with special characters, death of conifers (Greig and Redfern 1974). Where the fungus is present it normally enters healthy trees through roots that are in contact with infected stumps; then it may penetrate the heartwood and grow up into the stem causing decay, incipient decay, and stain. Pratt (1979) studied the damage done by the disease to Sitka spruce stands in Britain, varying in age from 23 to 50 years. He found that incipient decay occupied a greater volume of the stems of trees than either decay or stain and, although it could not readily be distinguished from sound wood, incipient decay significantly reduced the modulus of elasticity of the wood. More than half the sawn timber derived from 40- to 50-year-old trees of Sitka spruce comes from the lower 6–8 m of the stem, so rejection of butt logs from trees containing decayed and incipiently decayed wood causes serious losses in volume and value of the out-turn. The decay of roots also increases the susceptibility of conifer crops to windthrow (see page 34).

The spores of *H. annosum* are released from the fruit bodies throughout the year and are carried by wind to colonize freshly cut stumps, so clear cutting and thinning provide ideal conditions for its spread because infected stumps become strong sources of inoculum to invade neighbouring or succeeding crops. The spread of the fungus can be substantially reduced by treating freshly cut stumps with boron or urea, and pine stumps can also be inoculated with the competing fungus *Peniophora giganteum*. An infected site may be planted with a species showing some resistance to *H. annosum*, such as Corsican pine, Douglas fir, and Grand fir (Greig 1981), or the stumps of the old crop can be removed before the site is replanted (see page 75).

Pathogens of the mature stage

The onset of the mature stage is associated with decreasing height and diameter growth. The crowns of the trees broaden and branches become thicker. In regular stands that have been repeatedly thinned, the canopy is high and well-ventilated so attacks of leaf- and needle-infecting pathogens are more sporadic and dependent on climatic conditions than at earlier stages. Snow and wind breakage of crowns, death of branches caused by shading, fire and animal damage, felling grazes on stems, and wounds on roots following harvesting and road-making, all offer avenues of infection for decay fungi—which probably constitute the most important group of damaging fungi at this stage.

Insects

An individual stand in a forest may suffer damage from certain character-istic groups or species of insect pests at different times and stages of its growth. There will also be changes in the densities of the parasites, predators, and soil insects during the course of a rotation (Crooke 1979). The incidence and impact of insect pests are affected by several factors which include the following:

(1) the past history of the site and residues from the previous crop;
(2) insects carried on or near to the site by vehicles or machines which are harbouring pests of timber and bark, and residues from infested stands;
(3) climatic and other events which cause changes in the physiological state of the trees; and
(4) management decisions and forest operations.

Some examples follow to illustrate the action of these factors.

Insect pests of the establishment stage

The Large pine weevil, *Hylobius abietis*, is a pest of young conifers in Europe (Bevan 1987, p. 135). The main factor controlling numbers of this weevil is the presence of suitable breeding sites. Clear cuttings, and also heavy thinnings, fire, and windthrow provide abundant breeding sites under the bark of roots and stumps or the underside of logs or even large branches lying in direct contact with soil (Scott and King 1974), and the number of weevils rise. There is a spring peak of feeding by mature weevils emerging from hibernation, followed in June, July, and August by mass emergence of young weevils that have overwintered as larvae; the latter are very voracious and their feeding on the bark of young conifers does great damage. Among measures adopted to check the multiplication of the Large pine weevil are the formation of small coupes and the avoidance of fellings in adjoining areas except after a lapse of some years (page 67). This formation of small, segregated coupes is a useful precautionary measure, but it is effective only if accompanied by applications of insecticide to planting stock by dipping before planting and by spraying young plants on the forest site during June, July, and August.

The Black pine and spruce beetles of the genus *Hylastes* are among the most destructive pests of recently regenerated crops. These minute beetles breed in small roots, stumps, logs, and branches of conifers and the young adults feed during the spring and summer, mostly below ground level on the bark and cambium of young plants. Populations of bark beetles increase after clear cutting and outbreaks capable of doing very severe damage may occur when successive adjacent areas are clear felled. As in the case of the Large pine weevil, the formation of small segregated coupes is useful but this must be accompanied by applications of insecticide to plants before and after planting. According to Scott and King (1974) cultivation of the site by complete ploughing, ripping of the soil, or removal of stumps all reduce populations of *Hylastes* and *Hylobius*, but these treatments do not influence the numbers immigrating to the site.

Insect pests of the thicket stage

In the thicket stage sap-sucking insects, defoliators, and shoot borers are prominent pests. The Green spruce aphid *Elatobium abietinum* is a damaging sap-sucker causing loss of older needles in Sitka spruce; severe infestations can result in reduced growth increment. In Britain outbreaks begin during the thicket stage and can persist into later stages; they usually follow mild winters. In the forest, insecticidal control is impractical and undesirable, but one possible long-term approach is to produce cultivars with resistance to the pest (Bevan 1987, p. 135).

The Pine shoot moth *Rhyaciona buoliana* is a pest causing damage to the stems of Scots, Lodgepole, and Black pines in Europe. The larvae feed within shoots and buds in a zone 1–3 m above ground level, but small populations also persist at greater heights in older crops and can spread from these to neighbouring stands in the thicket stage. The commonest type of damage occurs when the terminal bud or shoot is killed but not all the side buds; this results in characteristic distortion of the stem, which often becomes the point of later breakage by wind and snow. Although species of parasites are plentiful they generally fail to exercise effective control. Silvicultural treatments which encourage rapid growth of young trees through the susceptible range of height will help to reduce attack by the Pine shoot moth and aid recovery (Scott 1972).

Insect pests of the pole stage

During the pole stage, when thinning begins, large-scale attacks by defoliators become possible. In the case of the Pine looper moth *Bupalus piniaria*, the likelihood of epidemics can be predicted from counts of over-wintering pupae in the soil, so insecticidal treatment can be applied, from the air, before serious damage is inflicted to Scots pine stands. Another possible control strategy is replacing Scots pine with Corsican pine or another resistant species.

Sawflies, such as *Gilpinia hercyniae* on spruces, and *Cephalcia lariciphila* on larch, can also reach damaging levels in the pole and later stages. The larvae of *G. hercyniae* are very susceptible to a naturally occurring but host-specific nuclear polyhedrosis virus, so effective control appears possible (Billany 1978). The prime cause of the collapse of populations of *C. lariciphila* in Britain has been the Ichneumon parasitoid *Olesicampe monticola* (Bevan 1987, p. 67).

Insect pests of the mature stage

As the mature stage approaches and is reached bark beetles and weevils become prominent. Endemic populations of the Pine shoot beetle *Tomicus piniperda* are kept high by regular fellings, windthrow, and fire damage. Trees weakened by drought or defoliation can be attacked and killed (Bevan 1987, p. 124). The adult beetles bore into and under the bark of felled pine logs or the bark of stems of unthrifty trees of breeding, but it is the damage caused by young and adult beetles when tunnelling in shoots during maturation and regeneration feeding which makes them important pests (Bevan 1962). Control requires de-barking after felling, or prompt removal of unbarked logs from the forest during the period from April to August and removal of winter-felled logs before breeding is accomplished. If unbarked logs must be left in the forest for longer than 6 weeks the stacks should be sprayed with insecticide.

Evans and King (1988) report that the Great spruce bark beetle *Dendroctonus micans* is regarded as 'a major pest of spruce from eastern Siberia to the west of Europe. It breeds under bark causing destruction of the cambium, which debilitates and, in extreme cases, kills the tree. Its reputation as a pest is based on the ability of female beetles to colonize living trees without the necessity of mass attack typical of most bark beetles.'

If a stand is currently free of *D. micans* the risk of an attack developing increases with proximity to infested stands, to extraction and haulage routes leading from infested stands, and to timber mills which may have converted infested timber. Stands of mature or over-mature trees in low rainfall areas which are growing rather poorly on soils prone to moisture deficits are susceptible to attack and this is made more likely by damage done when extracting timber or by windthrow. Control measures consist of sanitation fellings and biologial control. Infested trees are felled and the bark removed so that all stages of the beetle are disposed of. The peeled logs and the stumps are then sprayed with insecticide. Concerning biological control, this involves release and maintenance of the specific predator *Rhizophagus grandis*, which is regarded in Europe as an important means of containing populations of *D. micans* to an acceptable level.

Animals

The birds and animals of the forest depend on it for food and water, for shelter from wind, snow, and heat, for cover to help them escape from predators, and for space (Rochelle and Bunnell 1979). The regeneration, tending, and harvesting operations prescribed under the various silvicultural systems are disturbances which favour some animals and birds but may create problems for others.

Roe deer (*Capreolus capreolus*) is an animal that is favoured by the clear cutting system as it has been applied in some parts of Europe. Roe deer like clearings and forest edges for food and thickets for cover; they dislike crops in the pole stage and older because the understorey and vegetation are usually sparse. Roe deer move about 20 m away from the edge of a stand before starting to feed on grasses, herbs, and small tree seedlings. In addition they damage trees up to 20 years old by browsing the shoots and buds during winter and spring (Welch *et al.* 1988) and stripping off the bark in springtime (Chard 1970).

The trend towards small coupes of one-half to two hectares has reduced the distance between food and cover, while the tendency to shorten the rotation has raised the proportion of the whole forest that is under regeneration or in the thicket stage. Populations of Roe deer have risen greatly and the damage done by the grazing, browsing, and bark stripping has often reached unacceptable levels. As a consequence the number culled in the course of control measures has doubled since 1950 in Austria and the Karlsruhe area of Baden-Württemburg in West Germany (König and Gossow 1979).

Red deer (*Cervus elaphus*) obtain most of their food in young tree crops before the canopy closes and in clearings and glades within the forest (Ratcliffe 1985). Young trees may be browsed as the deer retire to cover and during winter when food is scarce. When browsing is severe growth increment is lost and stems become deformed. Bark stripping can also occur during late summer and winter and large wounds provide entry for microorganisms causing stain and rot, with consequent degrade of timber (page 46). Two other results of high population densities of Red deer are suppression or virtual extinction of the broad-leaved component in a conifer forest, and conflict with other land-users arising from marauding outwith the forest.

Although the amount of damage done by Red deer is broadly related to population density, simply reducing numbers is not sufficient to stop damage. Silviculture and management must be reconciled with the habitual pattern of dispersal and movement of animals within a forest (Chard 1966). When this is done an economic burden can often be transformed into a financial opportunity, as an example from Austria will show.

Jenkins and Reusz (1969) described the management of Red deer in a productive forest composed of Norway spruce (70 per cent), European larch (30 per cent) with some ash, hazel, and beech situated on north-facing slopes in the mountains of Steiermark. Alpine meadows and scrub lie above and agriculture is practised in the valleys below the main area of forest, the last two being separated by fencing. Red, roe and chamois (*Rupicapra rupicapra*) are all present in this part of Austria.

The silvicultural system is clear cutting with natural regeneration (see page 81), the size of individual coupes being limited to 2 ha. No felling is done in adjoining crops until the regeneration is 6 m tall. There is forage on the coupes while the young trees are becoming established, so about 160 ha of dispersed summer grazing is regularly available for Red deer. Snow lies over much of the area from November to March so natural winter feed is scarce and must be provided to prevent browsing, bark stripping, and also marauding into agricultural crops. The hefting behaviour of Red deer in winter is used to entice them into fenced enclosures in their own home valleys. These winter feeding and holding areas are 12–14 ha in extent. Each consists of mature forest and open meadow and holds groups of 30 to 40 deer of mixed ages and both sexes. Red deer are also fed on the alpine meadows near scrub woodland.

The accepted density of Red deer in Austrian forests is 1.5 animals/100 ha. The actual stocking is determined in autumn and the excess is selectively culled during the following spring to maintain a population of around 240 animals on the property that has been described. The management practised has greatly reduced winter damage to trees; the condition, weight, and quality of the deer is good; and the revenue from venison, stalking, and sale of live animals provides a small surplus for the owner.

Mechanical damage

The economic and social benefits of well-designed tools and machines are very great, but the manager must ensure that a tool or machine is suitable for a given task and that operators are trained in methods that are safe and do as little damage as possible to trees and site.

Regeneration can be damaged when seed bearers or reserves are felled. In the normal course of regeneration fellings, the young crop closes up and the results of damage disappear in a remarkably short time. In broad-leaved forests it is often customary after the final regeneration felling to cut back damaged saplings so as to stimulate regrowth of straight leading shoots; but if large seed trees or standards are felled into crops that have formed thicket, lasting damage can be done. The working rule expounded by Troup (1952) is 'don't leave the overwood any longer than is essential.' In some French forests the main branches of large seed bearers of oak and beech are lopped before felling—partly to prevent stems from splitting and partly to avoid damage to young regeneration.

In tropical moist forests the damage done to regeneration during felling and extraction of large old trees of the principal species can be considerable. The crowns of these trees may project over 200 to 300 m^2 of ground (Dawkins 1959) and natural saplings may have become quite tall. The need to control such damage has influenced the development of silvicultural systems in tropical moist forests of Africa and Asia (Wyatt-Smith 1987); this is discussed on page 156.

A major justification for thorough planning, close control, and careful execution of tending, felling, and harvesting operations is loss of timber value caused by stain and decay. Murray (1979) points out that the earlier in the rotation such injuries occur and the more pioneering the pathogens, the greater is the damage done. In spruce crops in Britain, the decay fungus *Stereum sanguinolentum* is an early colonizer, whereas reports from other countries state that *Fomes ignarius* enters wounds rather later. Stem wounds appear to be more readily infected by wood-decaying and staining fungi than root wounds. Pawsey and Gladman (1965) and El Atta and Hayes (1987) reported higher incidence of infection where large areas of sapwood were exposed, especially when the wood itself was damaged.

In pole stage and older crops, much mechanical damage can be done during thinnings because the space available for felling trees and extracting logs is limited. The extent of injury to the butts and superficial roots of standing trees depends on the resistance of the bark to abrasion, the diameter of the stems, the nature of the terrain, and the method of harvesting. The risk of damage to stems and roots rises with increasing slope and is also higher where an extraction path or rack changes direction or the ground becomes soft. Thus racks should be as straight as possible, be aligned up and down slopes, and be kept to dryer, firmer ground. They should also be widened at curves and at exits to main extraction routes. It may sometimes be necessary to protect particularly valuable trees with stakes.

On steep slopes ranging from 30 to 50 per cent in western North America and New Zealand, trees are felled across the slope (that is, along the contours) to reduce breakage of valuable logs. This directional felling is slower and requires more skill than felling up and down the slope, but a higher volume is recovered, the logs are presented more conveniently for extraction machines, and the fellers work under safer conditions (Murphy 1982). Similar results have been obtained from contour felling on steep slopes in Britain by Muhl (1987).

Some of the precautions commonly taken to reduce the extent and intensity of disturbance to the site during harvesting operations have already been mentioned in connexion with soil erosion, page 9 and protecting water quality, page 25. Whenever possible extraction is timed to coincide with periods when the bearing capacity of the soil is high. For example, in Scandinavia it is often done in winter when the ground is covered with snow. During felling on soils of low bearing capacity, such as peaty gleys, the logs are separated from the slash in such a way that the produce is correctly presented for extraction and the skidder or forwarder moves on a cushion of slash. Machines are being designed that exert low ground pressure and can be manoeuvred in restricted space. In this rapidly developing field the horse and other animals still retain a place.

Air pollution

The decline of the condition of some forests in central Europe (Bauer 1986) and eastern Canada, and the acidification of fresh waters in parts of

Scandinavia and Britain has caused much concern during the past 10 or 15 years. The deterioration in growth and health of Norway spruce, Scots pine, and beech appears to be particularly prevalent in parts of West Germany and Czechoslovakia and much research is being done to isolate the factors causing loss of foliage and general decline.

Damaging air pollution takes several forms. The most important gaseous pollutants are sulphur dioxide, nitrogen dioxide, ozone, and ammonia. High concentrations of atmospheric sulphur dioxide are damaging tree crops in the south-east of West Germany and Bohemia but may be less important now in Britain because of the action taken under the Clean Air Acts of 1954 and 1962.

Another major form of air pollution is that called 'acid rain'. It is defined by Innes (1987) as precipitation (rain, snow, and deposition of cloud water) that is more acidic than it would be in the absence of man-made pollutants. Some lakes and streams in Britain and Scandinavia have become acidified probably as the result of acidic deposition; this has also had a damaging effect on the flora and fauna. Water draining from some afforested catchments is more acidic than that draining from adjacent non-forested catchments. Although the reasons for this have not yet been fully established it is probable that interception of pollutants and natural salts by the canopy is the main cause. Treatment of forested areas on sensitive water catchments has been amended, notably by careful design of drainage systems (see page 25). Acidification of soils in some areas and possible effects of this are being closely studied.

The insidious action of air pollution can probably reduce the ability of trees to withstand the effects of adverse climatic and biotic influences, and the nature of this indirect damage is also being investigated in several countries. There is still much to be done before suitable silvicultural responses can be devised. Moreover, because the sources of air pollution and the sites where damage is done can be separated by long distances, concerted action to reduce pollutants and damage requires international co-operation and legislation acceptable to many countries.

5

The relation of silviculture to forest management

The complexity of modern forestry has stimulated publication of separate detailed accounts of silviculture and forest management, but Knüchel (1953), Dubordieu (1986b), Levack (1986), and many other authors have demonstrated the need to fuse the concepts of silviculture and management closely together. The purpose of this chapter is to summarize and comment on several topics which are treated in textbooks on forest management (see for example Johnston *et al.* 1967; Davis and Johnson 1987) from the viewpoint of silvicultural systems.

The role of forest management

The first task of forest managers is to determine the potential of a tract of land to produce the goods and services of a forest, and identify the main constraints on the practice of forestry. This information is used to define the productive, protective, and social functions of the forest which are set down, with their priorities, as the objects of management. The long-term objects are usually stated quite broadly, but the objects during planning periods of 10 to 20 years are given more precisely and they are reviewed and if need be, revised towards the end of each planning period.

When the objects of management are fixed the manager can begin his second task—that of planning how they will be achieved. Usually several possible methods will be available so the ecological, economic, and social criteria included in the objects of management are used to decide:

(1) which silvicultural system or systems are most suitable;
(2) how the forest is to be divided up for purposes of management and silviculture;
(3) how the yields of timber and other goods and services are to be regulated and sustained;
(4) how timber is to be harvested, processed and marketed;

(5)   how the forest is to be designed so that it functions efficiently and is as aesthetically pleasing as possible; and

(6)   how the forest enterprise is to be organized for the best use of its resources of land, labour, and capital.

The third task of forest management is to conduct the enterprise so that work and labour are in balance and financial targets are met. The quantities, costs, and revenues are brought together so that progress can be compared with the programmes of work prescribed in the plan of operations or working plan. If the planned targets prove unsuitable they are re-examined and modified until a satisfactory balance is achieved.

The division of the working plan area

The area covered by the plan of operations, called the working plan area, is divided into compartments. These are permanent management units for description, planning, and control, formed also to help forest staff, con-tractors, timber merchants, and others to find their way about the forest; so they must be clearly recognizable on the ground and marked on maps. The boundaries usually coincide with actual or projected road-lines and such natural features as ridges, plateau edges, rivers, streams, and valley bottoms. The size of compartments depends on terrain, intensity of working, and extent of the forest. The range commonly lies between 15 and 40 ha, small compartments being used in dissected terrain or where intensive forestry is practised over small areas of forest. Large compartments are typical of uniform terrain and more extensive working over considerable areas. In some tropical forests compartments may extend to 1 km^2 (see page 156).

Each compartment consists of one or several stands, which are sufficiently homogeneous in site, terrain, and tree crop to serve as units for silviculture and management. In the shelterwood uniform and selection systems of silviculture each compartment consists of one stand. The description of each stand records those characters that will affect its future response to silvicultural treatment, the quantity and quality of timber it will produce, how the trees will be harvested, and how logs will be processed into the final out-turn of sawn timber and other products. The stand description varies with circumstances but usually comprises:

1.   An assessment of land capability based on factors of climate, soil, and vegetation.

2.   Analyses of windthrow hazard (page 33), of terrain for use in harvest-ing, and of fire danger where needed (page 37).

3. Access from within the stand to the forest road network, and distances to markets.
4. Area, age class, species composition, structure, and the stage of development of the tree crop.
5. An assessment of the productivity of the tree crop. For pure, regular stands this usually is based on a top height/age relationship, called yield class or site index, linked to mensurational and management information including rotation (Philip 1983). For mixed and irregular stands productivity is often based on current annual increment.
6. Assessments of the health of the tree crop, including the occurrence of such defects as stem sweep or crookedness, resin pockets or decay which are likely to affect processing and final out-turn of products.

In several countries, including West Germany (Brünig 1980) and New Zealand (Levack 1986) stand models have been devised that allow simulation of the effects of alternative silvicultural treatments and management decisions on the growth, harvesting, and processing of a stand. The results may be presented in terms of volume, flows of cash, or both these.

Stands are allocated to working circles which are formed to satisfy a particular object of management and are worked under one silvicultural system with one set of prescriptions. A working circle may form all or part of the working plan area and the constituent stands may be concentrated in one part of the forest or be dispersed through it. Thus a working circle in which timber production is the prime function, and pure, regular stands are the best means of achieving it, can be distinguished from a working circle where site protection is the prime function and mixed, irregular stands are most suitable.

The sub-division of working circles for purposes of management and silviculture differs somewhat in each silvicultural system, as is described in the appropriate chapters. The main point to be made here is that the arrangement of stands into working circles is a decisive step towards shaping or designing the forest to satisfy the objects of management (see page 59).

Sustained yield and the normal growing stock

A sustained yield may be defined as a regular and continuing supply of the desired goods and services to the full capacity of the forest and without impairing the capability of the land. The concept of sustained yield can be applied to the conservation of a beautiful landscape, the provision of facilities for outdoor recreation or field sports, and to timber production, but we will concentrate here on the principal requirements for a sustained yield of timber. These are:

(1) the soil and growing stock of trees must be kept in a healthy and productive condition;

(2) the composition, structure, and stocking of each stand must match the capability of the site;

(3) the condition of the soil and increment of the growing stock must progressively be improved;

(4) the individual stands must be so arranged that they can be tended and harvested efficiently;

(5) within each unit of management the growing stock must have an appropriate distribution of size or age classes from youth to maturity; and

(6) reserves, in the form of trees or money or both, must be created as security against catastrophy.

The concept of sustained yield is central to all organized forestry but the interpretation of it varies with circumstances. In rural communities of Africa, Central and South America, and Asia which are isolated by poor roads and limited transport, loss of an assured and regular supply of fuel, poles, and other produce from nearby forests will wreck the whole economy of a village. Supplies can be restored and sustained with the aid of a simple silvicultural system, such as the coppice system, managed to create a growing stock that provides a steady annual sequence of maturing crops of a suitable species, usually with reserve stands to safeguard supplies. The closer this growing stock comes to satisfying the object of equal annual production the more ideal or 'normal' it is.

The private owner of a small forest will often aim at sustained yield based on a growing stock as close to normal in age or size classes as practicable, because many of the markets for his produce are quite close and he wishes to obtain an assured annual or periodic income from it; in addition he may wish to provide steady employment for local people. Again, where the primary functions of a forest are protective or social and timber production is an associated function, the creation of a near-normal growing stock can bring the desired benefits. Examples are forests where protection of the physical environment is the prime object (see page 23) and forests near towns or cities where outdoor recreation is the prime function (see page 182).

In forests managed primarily for timber production acceptance of the need for a normal growing stock varies greatly. Throughout Europe there are many fine examples of forests where the growing stock has been brought close to normal by lengthy application of a system of silviculture and management guided by well-defined and consistent objectives. Although many of these forests were begun in the late eighteenth and early nineteenth centuries, they have proved remarkably adaptable to changes in patterns of demand and the existence of a balanced growing stock has often reduced damage by wind, insect attack, disease and other hazards.

In modern industrial economies served by cheap and efficient transport there is often less need to make the growing stock of each forest 'normal'. Instead, several forests which have only a partial set of size or age classes can be grouped together into larger management units which have the full distribution of stands ranging from youth to maturity so that supplies are sustained from the whole group (Johnston *et al.* 1967). All the stands in the combined working plan area belong to one working circle, management is centralized, and much of the skilled work force moves around the several forests. Such arrangements are possible in large forest enterprises and when several owners co-operate together to supply a large processing plant.

There are two major developments which have induced a fundamental shift in attitude towards coupling sustained yield closely to a normal growing stock. The first of these is the formation of new forest resources based on pure, regular stands, often of exotic species; and the second is the use of research and development to devise ways of increasing the yield of forests (Matthews 1975).

Since the 1920s in Britain and New Zealand there has been extensive afforestation with spruces and Radiata pine, respectively. Formation of stands did not proceed at a steady pace but in three main periods of rapid planting interspersed with periods during which much less could be done. Timber production surged upwards as each major group of age classes became productive, and the mix of produce also changed from mainly small roundwood suitable for pulp and board mills to increasing proportions of saw logs in addition to the small roundwood. The aim in both countries, during the last 20 years, has been to sustain yield at progressively higher levels whilst avoiding the peaks and troughs due to the variable rate of planting. This has encouraged investment in new, large-scale processing plants and has also led to changes in silviculture and forest management. In New Zealand, the problem of reconciling the rate at which timber is produced in the forests to the rate at which it is taken up by the markets has been tackled by devising mathematical models of stands, working circles, and forests and using simulation (Allison 1980) and linear programming (Garcia 1986) to assist forest managers to find suitable solutions.

The second development bringing flexible attitudes to the normal growing stock as an essential element in sustained yield can be seen in Finland and Sweden. These are countries whose forests and forest industries provide

an important part of the national income. They have adopted large programmes of cultivation and drainage to raise the capability of forest soils.
The results of research into forest genetics and tree physiology are being
applied to improve the productive capacity of trees and crops, and utilize
more of the biomass produced. A range of new tools and machines has been
introduced, backed by comprehensive training facilities for the operators,
to improve regeneration, tending (including pruning of Scots pine), and
harvesting. The object of all this work is to raise the sustained yield of the
national forest estate so that supplies to large, integrated wood-using industries can be safeguarded. Conversion of the existing forests will take
many decades. Construction of mathematical models of stands and forests
for simulation and linear programming forms part of the process and a
fusing of silviculture and management is taking place.

Yield regulation

The prescribed yield is the force that drives the forest and its growing stock
toward the condition required by the objects of management (Köstler 1956,
p. 5). The annual or periodic yield is derived from thinnings, regeneration
fellings, and fellings of crops that have reached maturity. Several other
harvesting operations may be included in the prescribed yield. These may
include premature felling of crops damaged by wind and other agencies, or
those that are poorly adapted to the site conditions. Thinnings may favour
certain species so that a prescribed mixture is achieved, and a felling coupe
may mark the end of one management regime and the start of a new one. In
many forests in Britain, compartment boundaries are being modified after
felling to improve the efficiency and appearance of the forest by acting on
experience gained during the first rotation.

Where production of secure supplies of fuel wood and other small
produce is the object of management, as in the example given on page 52,
the annual cut is set so that the growing stock steadily evolves into a normal
series of coppice stands composed of a productive species, which produces
fuelwood with the desired qualities. Where the prime function of the forest
is protective the annual or periodic yield may be prescribed more flexibly so
that the varying needs of site protection can be satisfied.

Harvesting systems

The combination of methods used to fell trees, extract logs to roadside, and
haul them to market is called the harvesting system. The components of the

Figure 4. Harvesting systems. Bell Infield logger bunching in a pruned crop of Radiata pine. The trees were planted in patterns designed to improve access for extraction. New Zealand.

system include the tree crops, the road network, the vehicles, machines, and skilled workers available to the forest manager, and the markets for the produce (Conway 1982). The main constraints encountered when choosing a harvesting system are connected with climate, soil, and terrain and the need to protect the site and trees remaining on it. The system chosen is one that gives the lowest cost per cubic metre of timber delivered to market whilst causing the least possible damage to the site and growing stock. We have already referred on page 24 to methods for classifying terrain based on the bearing capacity of the soil, the obstacles on the ground, and the gradient and form of the slopes. Depth and hardness of snow cover are also included where this is a feature of the climate.

The produce moves from forest to market in several stages. The stage from stump to road is usually relatively short, seldom exceeding 1 km, but it is the most difficult and, on a weight basis, the most costly part of the harvesting system (Silversides 1981). Each silvicultural system produces a distinctive form of crop which generates particular problem in felling and extraction. There is continual improvement in laying out crops and extraction racks so that men, animals, or machines can be effectively and safely deployed in extracting logs by skidding, forwarding, the use of cable cranes, and other means (Fig. 4).

The advantages of simple methods and concentrated work for ease of management and control of costs are very great. Jones (1984) described harvesting equipment designed for a wide range of terrain and tree crops in Britain and some African countries. He stressed the need for simple harvesting systems with versatile and interchangeable equipment. Ease of servicing and facilities for training operators are also essential.

Planning and designing forest road networks

The network of roads, rides, extraction racks, and tending trails in a forest provides access for people, animals, vehicles, and machines engaged in regeneration, tending, and felling operations. It is also used for fire-fighting and other protective measures, and in harvesting and transporting produce to markets. Access by the general public may be provided by part of the road network. The details of the network required for each silvicultural system differs and these special features are described in the chapters on each system. Certain general principles governing design, standards of construction, and spacing of roads are presented here.

Roads designed and constructed to meet the needs of harvesting will almost certainly satisfy most if not all other needs (Rowan 1976). The investment in roads depends on several factors. The most important are the volume of timber produced by the forest, the kind of harvesting system or systems used, and the standard of road required for timber haulage. The costs incurred are the capital cost of construction and the recurrent costs of maintaining the road itself, the bridges, culverts, fords, passing and turning places, and the sites for stacking logs which may have to be built.

The standard of road construction is directly influenced by the method of haulage (Fig. 5). If timber is to be carried direct to the customer on large lorries from any part of the forest, high specification roads giving access in all weather will be required. If the timber is to be moved by tractor and trailer to a collecting point alongside a public road, an intermediate specification will meet the case. The roads built to high and intermediate specifications generally form the permanent element of the road network. Temporary roads or extraction paths laid out for tractors or specialized harvesting machines can be built to lower specifications; this part of the

Figure 5. Main extraction route on level ground. The beech stands have been produced under the uniform system. Lyons-la-Forêt, Normandy, France.

network tends to be more flexible, and as already noted, is modified as new harvesting methods and machines are developed.

Decisions on the method of extraction and standards of road construction together with an assessment of the volume of timber likely to be produced now and in the future leads to selection of road spacing (Rowan 1976). The actual length of road that has to be constructed to serve a given area exceeds the length, which calculations of road spacing indicate by amounts ranging from 25 to 35 per cent in lowland forests and favourable upland terrain to 45 per cent on difficult terrain. This is because roads are seldom straight. The need to conform to topographical features, the use of bends to gain height at acceptable gradients, and the limits on stream and river crossings all increase the length of road required to get from one point to another. Junctions and stacking sites, connections with public roads, rights of access, and the size and shape of forests also affect the length of road that must be built and maintained.

The detailed design of road alignments and cross-sections, and of bridges, passing places, and other works is a civil engineering task, as also are the precautions taken to prevent erosion and pollution of water courses.

Design which reduces the impact of harvesting systems and road networks on the landscape often receives greatest attention in areas of high landscape value, but if foresters wish to reduce recurrent costs and also keep the confidence of the general public, careful design should never be neglected.

The influence of timber markets on silviculture

The silvicultural treatment of managed forests depends greatly on the value in commerce of the timbers produced by the species concerned and on the specifications required by those who process and use the timbers. Hartig's version of the shelterwood uniform system applied to beech in the nineteenth century was designed to produce small diameter logs to meet very large demands for fuel wood; the version adopted nowadays in France and elsewhere (page 98) is designed to produce large diameter logs suitable for conversion to veneers.

In tropical moist forest successful application of silvicultural systems to convert them to sustained yield working has often followed closely on improved utilization of the very numerous component species. Until the 1930s only few species were felled and the logs were exported to markets overseas. As the indigenous populations increased, local demands for timber and fuel expanded and more species were harvested. This trend accelerated when the countries concerned began to process the timbers themselves rather than export logs. The number of species with wood properties and technical performance suitable for international markets remained small, but many more species were found to have properties suitable for wood-working on a factory scale, using industrial processes to produce panels, pulp, and paper for local markets. As a consequence modifications of the shelterwood uniform system have been successfully applied to some tropical forests and these are described in Chapter 12.

Principal timbers are those of major economic importance, because of their appearance, strength, stability, working properties, and natural durability. Their performance in service is very good and they command high prices in local and international markets. In mixed forests the species producing such timbers become those to which silviculture is primarily directed. Examples from mixed temperate forests are species of ash, hickory, oak, and walnut; and from tropical forests species of *Chlorophora, Entandrophragma, Gonostylus, Khaya, Lovoa, Pericopsis, Swietinia, Shorea, Tectona,* and *Triplochiton*.

The subsidiary or secondary species lack the intrinsic value or reputation of primary species in local or international markets. Some can be marketed because they are strong and durable, or can be processed easily on wood-working machines. Some species tend to be left in the forest by concessionaires so that their appearance in markets is irregular; others produce timbers of good quality but occur at low frequencies in the forest and this is the main reason for their subsidiary status. An important trend in marketing is to define the needs of the user and offer several species that will meet the specifications; this results in a steadier supply of timber with the desired characters.

The importance of producing a high proportion of first quality timber cannot be exaggerated. In this context 'quality' means closeness to the specifications of the market. Every user has his requirements and the silviculturist must try to meet them. To quote Bolton (1956), 'no industry of any sort whatever can prosper if it consistently floods the market with second or third rate material. First quality timber can always command a fair price, while poor quality timber, except in times of great scarcity, is—and rightly so—a drug on the market.'

The operations that significantly influence timber quality are those that improve the site and act on the crop as a whole, namely cultivation, drainage, irrigation, and applications of fertilizer; and those that act directly on selected trees influencing their growth, i.e. spacing distance and pattern, thinning, and pruning. However the influence of these operations on timber quality is indirect, and the process can be summarized thus (Brazier 1977):

Tree growth influences wood properties which affect the technical performance of wood and this in turn determines the acceptance of timber for various end uses.

The design of forests

The word 'design' has many meanings, most of which are applicable to forestry (Davis 1976). Design is the visual expression of the ideas that give purpose and direction to a forest enterprise. To be effective careful design must permeate throughout the enterprise so that the forest satisfies its productive, protective, and social functions as they are defined in the objects of management (Fig. 6). The purpose here is briefly to describe and discuss some aspects of forest design, emphasizing those affecting silvicultural systems.

The components of forest design are:

1. The underlying geology and surface features which together determine the character of the landscape.

Figure 6.   Clear cutting system. Two well-designed coupes on moderate slopes in hilly country. The main extraction routes are in the valley. *Crytomeria japonica*, Japan.

2.   The silvicultural system itself, which determines the method used to regenerate individual crops, the form of crop produced, and the arrangement of crops over the whole forest.
3.   The network of roads, rides, and racks or paths which link the whole forest together into a working unit.
4.   The external and internal boundaries and the treatment of these to protect the growing stock from damage, conserve wild plants and animals, and enhance amenity.

To quote Davis (1976 p. 101), 'design includes creativity, form, structure, purpose, organization of parts, and application to means of an end—artistry in a full sense. It includes design of structures and of land forms, and their combination into a pleasing and functional whole. Design implies imagination, innovation, and the expression and integration of ideas in some harmonious fashion to a desired end.'

The social functions of managed forests

The enjoyment of recreation in managed forests can be a natural result of sympathetic design. The extent to which opportunities are provided for different forms of recreation depends on the position of the forest relative to towns and cities, and on the objects of management.

The most popular and compatible forms of recreation in forests are those described as informal, that is, they depend for their appeal on natural factors. Informal recreation is an important function of forests near towns and cities and, where demand is strong, the silvicultural system chosen is often one that is 'natural' and causes least disruption of well-loved views and features of the landscape. Mixed stands with irregular structure, long rotations, and open spaces are common features of forests used for informal recreation.

Those who enjoy camping and caravanning prefer sheltered and flat or slightly undulating sites which are well-drained and near to a stream, river, or lake. Careful design of sites and the supporting services is essential if the people using them are not unwittingly to destroy the habitat whilst they are enjoying outdoor recreation.

Rights of user

The existence of rights of user, particularly as it affects the kind of produce to be provided, may influence the choice of a silvicultural system. In the case of grazing rights, definite areas have to be closed for regeneration. As already noted on page 9, systems with short regeneration periods and the clear cutting system are preferable to long-period systems and the selection system.

It is rare to find areas of tropical forest in which forest dwellers do not claim to have tribal ownership or rights to carry out various operations for their own use—small clearings for agriculture, felling trees for timber, collecting minor forest produce, fishing, and hunting. In most cases these rights are, or should be satisfactorily resolved when an area is established as part of the permanent forest estate.

The broad scope of forestry

A forester is an ecologist, manager of resources and people, and business man alert to the needs of markets. One definition of the broad scope of forestry reads as follows (Science Council of Canada 1973):

'Forestry is the science, business and art of managing and conserving forests and associated lands for continuing economic, social, and environmental benefit. It involves the balanced management of forest resources for optimum yields of wood products, abundant wildlife, plentiful supplies of pure water, attractive scenic and recreational environments in wild, rural, and urban settings, and a variety of other services and products. Forestry draws on knowledge and experience from many other disciplines and other professions.'

One aspect of the broad scope of forestry that has become prominent, especially in tropical countries, is that included in the collective term 'agro-forestry'. This term embraces land use systems and practices in which trees and shrubs are grown together with agricultural crops and with domestic animals. There are a great variety of traditional and relatively new agro-forestry systems. Some of these are described at various points in the chapters which follow, because they originated in different ways and at different times during the evolution of silvicultural systems. Discussion of agro-forestry systems in general appears in Chapter 20. They are mentioned here to alert the reader to an important trend in which some of the productive, protective, and social functions of forests are being met by systems combining agriculture and silviculture in various ways.

Attempts to encourage more widespread use of 'agro-forestry' systems and practices must be supported by appropriate schemes of management which are based on detailed studies of the social and economic condition of rural people. This important task has been given the name 'social forestry' which seeks to involve indigenous people in rural development programmes.

PART II

Silvicultural systems in practice

6

The clear cutting system

Fr. Coupe rase, coupe à blanc; Ger. Kahlschlag, kahlhieb; Span. Corta a hecho

General description

Under this system successive areas are clear felled and regenerated, most frequently by artificial means but sometimes naturally. As a general rule the coupe should be completely cleared, although pre-existing poles and saplings may be left if they occur in promising groups large enough to form self-contained crops. Under ordinary conditions, however, isolated poles and saplings should be removed, as these are likely to develop into branchy trees as well as interfere with the new crop.

In special cases it may be necessary to introduce temporary nurses to assist in the establishment of the new crop. In the tropics unmarketable trees are sometimes killed by girdling and applying a herbicide in order to save the cost of felling them.

All timber and other marketable material should be removed at once, and if the slash remaining is likely to interfere seriously with sowing or planting, or with the establishment of the new crop, it should be burnt or macerated (see page 16).

Size, form, and arrangement of coupes

In its ideal form the clear cutting system involves felling and regenerating each year equal areas called coupes on which the stands have reached the pre-determined age of maturity or rotation. If this process is repeated each year for the whole rotation and no accidents have happened to interrupt the continuity, a normal series of gradations aged 1,2,3 ... *r* years (*r* being the number of years in the rotation) will have been established. Where the productivity of the soil varies appreciably from place to place, coupes not equal in area but which are equally productive are formed, those on the poorer soils being larger than those on more fertile soils.

The ideal form of the clear cutting system, involving felling and regenerating equal annual coupes and the formation of an uninterrupted series of age gradations, has been found impracticable in high forest except under very favourable conditions. Modelled on the coppice system (page 190), it was introduced in the coniferous forests of Germany at the end of the eighteenth century but proved a failure owing to the impossibility of maintaining the ideal regularity aimed at; the longer the rotation, the more difficult it became to fulfil the aim. For this reason the modification of periodic instead of annual coupes was adopted at the beginning of the nineteenth century. Under this modification it may be arranged, for instance, to fell and regenerate an area ten times that of the theoretical annual coupe in a period of 10 years, the actual area felled annually varying from year to year within this period. Such a procedure is indicated where considerable latitude is desired for economic reasons, or where natural regeneration from adjoining woods is relied on, or where artificial regeneration is carried out but supplies of seed or plants are obtainable only at irregular intervals.

The size, form, and arrangements of the coupes vary greatly according to local conditions and requirements. Where damage from cold, dry, or strong winds, frost, drought, insects, disease, erosion of hillsides, or other dangers is not to be feared the coupes may be of any size or shape compatible with local topography and amenity, harvesting systems, and other considerations (Fig. 6, page 60); but in certain cases special measures may be necessary.

A protective measure sometimes adopted against cold winds, frost, snow, drought, and other dangers, is to make within a mature crop small scattered coupes which are usually regenerated artificially. Further coupes are made from time to time in the remaining portions of the mature crop until the whole area has been felled and regenerated. As a means of introducing sensitive species or of producing somewhat uneven-aged crops, this procedure has its merits, but it is difficult to control and is expensive to operate owing to the scattered nature of the fellings; it also increases the danger from wind, while on steep hillsides damage may be caused to groups of young growth during the extraction of timber. Hence where protective measures are required it is preferable to adopt systematic clear cutting procedures designed to prevent damage by extremes of climate, insects, diseases, and the extraction of timber. It is also necessary to forestall possible effects of clear cutting on soil erosion, and the hydrology and amenity of the area (page 13).

Protection against wind

In regions liable to gales it is important to avoid clear cuttings that will suddenly expose the interior of a stand to strong winds and the coupes should be so arranged that fellings in successive areas proceed against the prevailing gale direction. In the British Isles and in north-west Europe the prevailing direction of gales is from the west and south-west so fellings proceed from east to west or from north-east to south-west, with local diversions in hilly country to allow for modifications of wind direction by the terrain. By this arrangement there is always a sheltering stand, or failing this there should always be a good shelter belt, to windward of the felled coupe. The usual procedure therefore, is to select areas containing crops that will be felled for a limited period ahead, say 20 years, and arrange the coupes in those areas in such a way that fellings proceed against the wind direction. A further precaution used in the mountains of central Europe is to form narrow coupes, approximating to cleared strips, running at right angles to the gale direction, and to avoid large clearings. In the Norway spruce forests of Saxony in East Germany these coupes vary from about 20 to 100 m or more in width.

When considering how to arrange narrow coupes to provide protection against sun, frost, and cold winds, there is an overlap with the strip systems, which are described in Chapter 11. This subject is dealt with in detail there.

Protection against insect attack and animal damage

Precautions against the risk of damage by certain insects may also influence the arrangements of coupes. Among measures to check the multiplication of the Large pine weevil (*Hylobius abietis*) and the Black pine beetles of the genus *Hylastes*, are the formation of small coupes and the avoidance of fellings in contiguous areas except after the lapse of some years. However, as explained on page 41, the formation of small, segregated coupes is effective only if accompanied by applications of insecticide to young regeneration. Concerning animal damage, in the clear cutting system the coupes are usually adjacent to pole stage and older crops and when the coupes are small or long and narrow, the length of the edges between young and older crops per unit area of forest is high. The converse is also true when the coupes are larger, so size and shape of coupe often affects the size of wildlife populations and the amount of damage done to young trees by grazing and browsing animals (König and Gossow 1979).

Cutting sections

In order to faciliate the formation of small or narrow coupes and their arrangement in some regular order, it is usual to divide each main unit of the forest into a number of cutting sections. These may be defined as units,

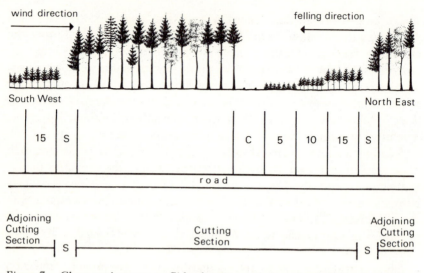

Figure 7.   Clear cutting system. Side elevation and plan of a single cutting section. S,S, severance cuttings; C, coupe of the year. Numbers denote ages of crops.

usually of limited size, formed for some special purpose such as protection against damage by wind or insects. Within each cutting section the coupes are arranged to meet the requirements of the case. Figure 7 shows how coupes may be arranged in cutting sections so that the general direction of felling is from north-east to south-west against the prevailing direction of gales, while at the same time intervals of 5 years are allowed for between successive fellings in adjacent coupes. Narrow coupes arranged in this way also serve as a protection against frost and cold north-east winds (page 141). Wherever possible, cutting sections should be bounded on the windward side by shelterbelts treated in the manner described on page 35. Where permanent cutting sections are laid out, roads and rides bordered by such belts make the best boundaries.

In the formation of cutting sections it may be necessary to divide existing forests into two or more parts. If this can be foreseen long enough ahead, shelterbelts may be produced by making severance cuttings. These are cleared lines of varying width, usually 10–15 m, cut through the forest while it is still young in order to induce low branching and marked stem taper in the edge trees. Severance cuttings which run more or less at right angles to the prevailing wind direction may be used as roads or rides, or they may be planted up.

To simplify extraction of timber, where the terrain allows, the cutting sections should be bounded by roads running along the narrow ends of the coupes; this is a common arrangement on more or less level ground, where the main extraction routes run parallel to each other in a general north-east to south-west direction.

The length of a cutting section may vary considerably. In the Tharandt school forest near Dresden in East Germany, cutting sections generally extend over two compartments and have an average length of about 600 m. The general tendency, however, is to decrease the size of cutting sections in order to provide flexibility in their working.

Form and arrangement of coupes in hilly country

In hilly country, considerations of protection against wind and insects must be combined with measures to prevent damage to young crops during extraction of timber. As already noted on page 35, in hilly terrain the prevailing winds tend to be diverted according to the trend of the valleys, and the local wind direction has to be determined. The coupes themselves are generally, but not always, long and narrow and are arranged either: (1) with the long axis running up and down the slope and the fellings proceeding against the local gale direction, and with a road running along the lower edge of the cutting section (see Fig. 8); or (2) with the long axis running more or less contour-wise, but slightly inclined to the horizontal, if this is necessary to avoid exposing unprotected stands to the prevailing

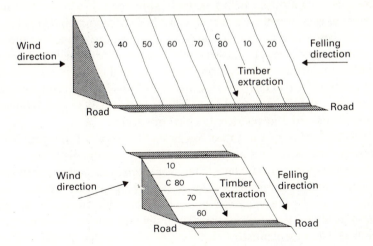

Figure 8. Clear cutting system in hilly country showing alternative arrangements of coupes. In practice the boundaries would be shaped so as to fit the landscape. Numbers denote ages of crops. C, coupe of the year.

winds. The felling direction is downhill (Fig. 8). Under the latter arrangement short cutting sections are necessary, bounded by roads at fairly frequent intervals so that large timber does not have to be extracted through young crops.

Design of coupes in relation to the landscape

The visual impact of the clear cutting system can be made acceptable by considering the size, form, and arrangement of coupes in relation to the visual characters of the landscape (page 28). Kielder forest in northern England lies at elevations between 150 and 500 m in an extensive area of rolling hills dissected by valleys which contain the watercourses. The slopes are moderate, but locally steep and exposure to wind is severe on sites above 300 m. The forest consists of 40 000 ha, mainly of spruce planted on gleyed and peaty soils from 1926, but mainly from 1946 to 1960. In 1982 a large reservoir, called Kielder Water, was completed in the middle of the forest and this has become a popular site for outdoor recreation. The water level is at 190 m.

The forest is managed using the clear cutting system. Felling and replanting of second rotation crops has been in progress since the 1970s. Three zones, mainly based on elevation, are now recognized for purposes of management and these correlate broadly with site capability, windthrow hazard, silvicultural treatments, and recreational use by the public (Hibberd 1985). The shape of the felling coupes (which equate with compartments and stands) is approximately rectangular, the vertical margins following the water courses while the upper and lower margins are formed by the main extraction routes (Fig. 9). These coupes blend well with the broad sweep of the hills, the detailed visual interest being provided by the valleys with their water courses. Size of coupe varies from 5 to 25 ha in the lowest zone, 25 to 50 ha in the middle zone, and 50 to 100 ha in the highest zone with severe windthrow hazard and restricted choice of species.

This structuring of the forest has brought several advantages. A long-term sequence of fellings has been established and the distribution of age classes is being improved. Harvesting, restocking, and other operations can be more readily organized. Roe deer (*Capreolus capreolus*) which caused much damage in the past, are now under closer control so that native broad-leaved species are appearing naturally and can be supplemented by more broad-leaves planted in small steep-sided valleys. In the long term, this large and productive forest should become more resistant to wind damage. It will also provide a varied and attractive background to recreation on Kielder Water (Fig. 9).

Figure 9. Clear cutting system with artificial regeneration. Large coupes with boundaries adjusted to features of the landscape. A holiday chalet is on the left, looking over Kielder Water. Kielder forest, Northumberland, Britain.

Clear cutting with artificial regeneration

Under the clear cutting system artificial regeneration is far more commonly practised than natural regeneration, planting being more general than direct seeding. However, aerial seeding of Jack pine, Douglas fir, and species of *Eucalyptus* is done in North America and Australia, respectively, while direct sowing of acorns of oak is still practised in several parts of Europe.

As a rule it is important that the site should be prepared for direct seeding or planting and regenerated as soon after felling as possible because cleared areas left for a year or more tend to become covered with weeds and regrowth of woody species, and the condition of the soil deteriorates, dry soils becoming drier on the surface and wet soils becoming more swampy.

During the early years before canopy closure, the young trees depend on nutrients supplied by the soil and they must compete with the ground vegetation for the light, water, and nutrients that are available on the site. Treatment of slash, cultivation and drainage of the soil, control of competing weeds, application of fertilizers to supply nutrient elements, and irrigation are used as required to create conditions that favour rapid establishment of the new crop.

Treatment of slash

Where the quantity of slash is not great and it does not constitute a fire hazard or cause damaging insects to increase it can be left to break down slowly; the release of nutrients and the improvement of the micro-climate near the ground can aid young plants to survive and grow. In some cases, for instance in parts of Scandinavia, the slash is incorporated by machine into the surface layers of the soil in mixture with the humus. Sometimes the slash is cut up by hand or macerated by machine and distributed over the soil surface to act as a mulch; this technique is used in dry regions to suppress weeds, conserve moisture, reduce the impact of sun, rain, and wind on the surface of the soil and enhance microbial activity thus releasing nutrients. Where the quantity of slash is large so that it constitutes a fire hazard or encourages the increase of bark beetles and weevils it must be reduced in volume or destroyed. The most common method is to burn the slash, and controlled burning has been particularly well-developed in those regions where the climate includes a dry season (page 16).

Ground cultivation and drainage

Scarifiers are machines for shallow cultivation. They are used, particularly in Scandinavia, to encourage natural regeneration, and to improve the survival and early growth of seedlings and transplants on felled coupes. Scarifiers are a useful alternative to ploughs where the amount of ground vegetation is limited and part of the old drainage system has persisted together with root channels. On wetter sites, scarifiers produce a raised, aerated, and weed-free planting site with approximate dimensions 50 × 60 × 30 cm. On dryer sites the machines push the slash aside and expose patches of mineral soil with discs or mattocks.

There are many sites where deeper cultivation will improve the aeration, drainage, temperature regime, and nutrient status of the soil thus encouraging the growth of tree roots—which in turn increases the growth of leaves and shoots, the subsequent yield of timber, and the resistance of trees to windthrow (Thompson 1984). Ploughing also inverts the soil to form ridges; this inversion of the soil smothers the ground vegetation, which in turn releases nutrients as it decays, and the broken surface of the ridge provides seed beds or places to plant seedlings or transplants. When restocking clear cut coupes the roots of the previous crop may be left undisturbed, or be dislodged (Low 1985), or even completely extracted (page 75).

The technique of cultivation varies with the type of soil. On freely draining soils, such as brown earths and podsols, a plough fitted with a tine or subsoiler and double mouldboards will mix humus and soil, suppress vegetation, and create places for seeds or plants. On impeded soils with a distinct layer that is compacted, cemented, or indurated the shattering action of a tine plough is used to fragment the layer and increase the volume of rootable soil. And on impervious soils, which include gleys and peats, drainage ploughs are used to remove as much water as possible from the site and encourage development of a broad platform of roots (Savill and Evans 1986).

In the south-west of England where plantations of Sitka spruce are grown on sites with surface-water gley soils and high windthrow hazard, another technique has been adopted when regenerating clear cut coupes. The potential of these productive sites for growing timber is high but the strong growth of *Molinia* grass, rushes (*Juncus* spp.), and broad-leaved shrubs must be controlled, the drainage must be improved, and the future stability of the tree crop assured. Deep drains are dug by tracked excavator and the spoil is distributed in large mounds at 2.1 to 2.6 m square spacing. The mounds are allowed to settle for 9 months and transplants of Sitka spruce can then be planted on well-drained positions which are raised above the ground water and remain free of weeds; phosphate fertilizer is also added at time of planting and directly benefits the young trees. The early growth of the new crop is rapid and the root systems quickly develop a radial pattern.

Ploughing accelerates run-off and may lead to erosion and increased amounts of sediment in streams (Mills 1980). Cutting cross-drains will greatly reduce this effect. It is also prudent to stop plough furrows short of streams and lakes (page 25). Where aesthetic considerations are important, the visual impact of ploughing can be reduced by adjusting the directions of the ridges and furrows so that they do not produce visual conflicts with the curves of prominent features of the landscape.

Weed control

When establishing the new crop, control of competing weeds is usually essential; its effectiveness can be increased and costs reduced by closely integrating treatments to control weeds with those used to dispose of slash and cultivate the soil. For instance, the soil disturbance caused by deep complete ploughing will often delay the regrowth of weeds.

A useful approach to weed control is described by Crowther (1976) who classified the weeds commonly encountered in Britain by their increasing competitive power, namely, soft and fine grasses, herbaceous broad-leaved weeds and grasses, coarse grasses and rushes (*Juncus* spp.), bracken (*Pteris aquilinum*), heather (*Calluna vulgaris*), and broad-leaved trees and shrubs of coppice or seedling origin. Each class is subjected to appropriate manual, mechanical, and chemical treatment, either singly or in combination. In many cases chemical herbicides, applied with appropriate safeguards, offer the most effective and cheapest method, often because one application will have a longer-lasting effect than hand or machine weeding. Techniques for applying herbicides which reduce the volume and weight of active substance and diluent that must be carried into the forest are continually being developed.

Supplementary nutrition

When fertilizer is applied the nutrient element becomes distributed between the trees, ground vegetation, the layer of litter and humus, and the mineral soil; a small portion is lost from the site through leaching or, in the case of nitrogen, by gaseous diffusion. If the trees are deficient in the nutrient element applied they will respond by growing at a quicker rate for a period, and the enhanced growth appears mainly to be related to the amount of nutrient that has accumulated in the foliage. So it is important not to allow the crowns to become too sparse before applying fertilizers. Miller (1981) suggested that it might be desirable to apply once to build up the mass of foliage and a second time to fill the leaves with the desired nutrient.

The requirements for fertilizer alter with the stage of development of the tree crop. Attiwill (1979) and Miller (1981) recognized three different stages in the life of a tree crop, two of which are important here. In the establishment stage the developing crowns of the young trees require large amounts of all nutrients, but at this time much of the soil volume is occupied by the roots of competing vegetation; moreover the weeds intercept a large part of the nutrients carried in rain-water. The concentration of available nutrients in the soil is the critical factor controlling the rate of crown development, and responses to a wide range of nutrients applied in fertilizers can be expected.

The second stage of development is reached when the crop has formed a complete canopy. The demand for nutrients is high but now there is cycling within the trees and the ecosystem; nutrients carried in rain and aerosols are also intercepted by the canopy of leaves and the dense network of shallow roots with their mycorrhizal hyphae. For these reasons the demands made on the nutrients in the soil are low and responses to fertilizers are unlikely, unless the crop is thinned or damaged by insect attack and the foliage that is lost must be replaced.

It will be seen that fertilizers are most likely to be of value prior to canopy closure and they are often applied at or soon after planting or direct seeding. The future development of a crop can be greatly influenced by early corrections of deficiencies and those are indicated in general terms by the composition of the vegetation, by soil type within geographic regions defined by elevation and latitude, by visual symptoms of the foliage, and more precisely by chemical analysis of the foliage. Critical and optimal levels of major and minor nutrients should be determined by experiment to fix the rates and forms of fertilizer that should be applied.

Special techniques for artificial regeneration

On some of the high moors of Bohemia in Czechoslovakia, Norway spruce regenerated artificially on clear cut coupes cannot withstand the severe frosts without the aid of temporary nurses, such as birch and rowan, which appear naturally or are introduced artificially. The use of temporary nurses has also been useful, mainly in eastern parts of Britain, when planting exotics such as Douglas fir and Grand fir on sites prone to late spring frosts (see page 235). However Wessely (1853) stated that in the Austrian mountains some of the finest young stands of beech (which is a tender species) were produced on clear cut areas, and Anderson (1949) reported that the fine mature beech in the Forêt de Soignes near Brussels in Belgium, had also been established on clear cut areas.

In Thetford forest in east England almost half the trees in young crops of Scots pine were being killed by *Heterobasidion annosum*, the most severe damage occurring on land that had been abandoned by agriculture (Greig and Low 1975) where the soil was derived from drift of chalk and sand and had a pH exceeding 6.0. The tree crops most affected are being felled at 45 years and, to prevent damage by *Heterobasidion* to the new crops, the coupes are cleared by a tracked excavator fitted with a large spike which lifts all the stumps out of the ground; they are then raked into windrows 40 m apart by a tractor with powered fork-rake. The light soil and flat terrain permits replanting by machine using seedlings of Corsican pine in containers (Winterflood 1976). The cost of this work is amply covered by the higher yield of timber obtained from a healthy new crop.

Artificial regeneration by direct seeding

The conditions for successful regeneration by direct seeding are:

(1)   an abundant supply of viable seed of a suitable provenance;
(2)   creation of seed-beds favourable to the rapid germination of seed;
(3)   efficient dispersal of seed when conditions of climate and site are favourable;
(4)   a low rate of seed predation by birds, animals, and insects; and,
(5)   protection of the young seedlings from excessive competition and damage during the first season of growth.

These are closely similar to the conditions required for successful natural regeneration (page 92).

The species suited to direct seeding generally have seed that is small, like *Eucalyptus*, or of medium size, like many pines. (The acorns of oak are an exception to this rule.) The seed must germinate readily, if necessary on the soil surface because only light covering is normally possible, and the early growth of the shoot and root systems must be rapid. In general, light-demanding pioneer species are the most suitable, especially those that regenerate after wild fires or disturbance of the soil.

Direct seeding requires a supply of viable seed up to ten times greater than that used to raise sufficient plants in a nursery to regenerate a coupe of the same area. Thus it is most suited for species which produce good or very good crops at frequent intervals, or which can be kept in cold storage for several years to provide continuity of supply. Douglas fir produces good or very good crops at intervals of 5 to 7 years but it can be stored for 5 or more years. Seed of provenances that are suited to the climate and site con-ditions and have the required rate of growth, stem form, and other desired characters is often produced in special seed production areas that have been thinned and fertilized to increase the quantity and quality of the seed. So far, seed orchards formed to produce seed of superior cultivars are not sufficient in area or age to produce the amounts of seed needed for direct seeding.

In general, the most favourable seed-bed for germination of seed consists of mineral soil into which some humus has been thoroughly mixed. Survival of seedlings is greatest when the competition from vegetation is weak, moisture is plentiful, and the young plants are protected from extremely high air temperatures near to the ground; the latter condition can be met by the shade of some slash or by the leaves of herbaceous plants. It is not necessary or desirable to expose large continuous areas of soil to receive the seed, so appropriate techniques have been devised for different conditions of vegetative cover, ground roughness, and soil type. The disturbance of soil during harvesting often creates sufficient seed-beds, especially if ground skidding is used to extract the logs.

Seed can be sown by hand when the coupe is small in area and irregular in shape, but the conditions presented by the roughness of the ground, surface vegetation, and slash are not severe. Patches 30 to 60 cm across are cultivated and a fixed number of seeds, usually six to ten, is sown and gently pressed into the soil. The patches are spaced at, say, 2 × 2 m to give up to 2500/ha. Under favourable soil conditions a light tool is used, combining cultivator and seed sower fitted with a metering device, and it is possible to direct seed at the rate of 1–2 ha/day. Another technique being tried in Israel, Scandinavia, and Canada is to cover the seeds with cone or cup-shaped shelters, usually made of plastic (Solbrana 1982; Sahlen 1984).

On larger coupes where the bearing capacity of the soil, surface roughness, and slope allow the passage of wheeled tractors, the soil is cultivated by machine to form seed-beds in strips, ridges, or furrows 40–70 m wide spaced at the desired distance apart. The seed is sown broadcast or in drills. The cultivator may have a seed sowing attachment or the seed can be sown by a cyclone blower held in the hand.

When the coupes are large and in remote mountainous areas, where the ground may be rough or very rough and the slopes steep, direct sowing by aircraft is now widely practised, usually in combination with controlled broadcast burning to prepare the seed-bed. In the *Eucalyptus* forests of Australia and especially in mountainous parts of Victoria, New South Wales and Tasmania, periodic wild fires caused by man or lightning are often followed by natural regeneration, which commonly appears where the mineral soil has been exposed and is strewn with ashes. The clear cutting system with broadcast burning followed by direct seeding from the air is used to regenerate stands of *E. regnans* and *E. delegatensis* and 8000 to 10 000 ha are being treated each year (National Academy of Sciences 1981). These two species are strong light demanders. Large coupes are clear cut, the slash and ground cover are burnt, and seed is scattered over the seed-bed from aircraft. The seed is tested to ensure high viability and is treated to reduce predation with an external coating of insecticide and fungicide bound to the seed coat with kaolin; dye is also added to make the treated seed easier to see so that the pilot and ground crew can check that it is being evenly distributed. The combination of weather conditions in which it is safe to burn slash and suitable to sow seed occur on only a few days during each year. If these favourable conditions are not exploited the regrowth of scrub species and loss of slash by decay make it difficult to achieve the hot fire needed to expose the mineral soil and create a good seed-bed.

Much of the direct aerial seeding in North America has been done with species of pine and Douglas fir. In western Oregon temperatures close to the surface of the soil on newly burned coupes can reach 60°C in summer—too high for successful direct seeding. Seed of mustard (*Brassica juncea*) has been sown at a low rate with the tree seed; it sprouts quickly, reduces soil erosion and soil temperatures, and provides shade for the Douglas fir seedlings (McKell and Finnis 1957).

When Douglas fir seed begins to fall it is eaten by mice and shrews (Fowells 1965) and sharp increases in seed-eating birds have been observed on newly clear cut coupes. Direct seeding is only practical if the losses due to predators can be reduced, so it is essential to know the species and assess the numbers of mice and other seed-eating animals, birds, and insects before attempting to apply the method. A representative sample of small areas of seed-bed should be sown with viable seed and covered with caps of varying sizes of mesh selected to keep out potential predators. Having identified the most destructive, action can be taken to reduce their effect. Direct seeding can be done at a season when the numbers of a predator are low; preparation of the seed-bed by controlled burning or scarification will also reduce populations of seed predators; or the seed can be coated with repellents.

The advantages of direct seeding are:

1. It can be organized relatively easily and requires relatively few people so it is especially suited to large coupes in sparsely populated regions with rough terrain and difficult access.
2. A complex infrastructure of nurseries, roads, and transport to supply plants and housing for workers in remote areas is not required, hence capital outlays are low.
3. If direct seeding from the air is used, the aircraft can often be obtained on short-term contract from agricultural aviation services or sometimes from an air force.
4. If successful, direct seeding is cheaper than planting and most money is saved on sites where the costs of planting are usually high.

The disadvantages of direct seeding are:

1. Seed collection equipment and seed storage facilities are needed to sustain the large amounts of seed that are required.
2. It is restricted to relatively few species and the conjunction of favourable climatic and site conditions is often critical and fleeting. Prompt

action is required to obtain good germination and survival and the risk of failure is higher than with planting.

3. Some of the chemicals used to coat the seed and reduce predation are dangerous to man, birds, and animals. They must be carefully used and controlled.

Artificial regeneration with the aid of field crops

Fr. Cultures intercalaires, culture sylvicole et agricole; Ger. Waldfeldbau

Under this special form of artificial regeneration, field crops are cultivated temporarily on the cleared coupes for one or more years, the sowing or planting of the tree species being done before, along with, or after the sowing of the agricultural crop. This practice, which varied greatly in detail, was common in Europe until the nineteenth century but is now virtually obsolete there. In the tropics it has grown greatly in importance.

In the tropics there is a more or less unregulated method of cyclic cultivation whereby cultivators cut some or all of the tree crop during the period of least rainfall, burn it shortly before the rains, and raise agricultural crops for one or more years before moving on to another site and repeating the process. Secondary forest colonizes the abandoned sites and the cultivators return after intervals of 4 to 20 years. This shifting cultivation as it is called, is very widely practised (Sanchez 1976), and has numerous native names, some of the better known being *chena* in Sri Lanka, *conuco* in South America, *kiangin* in the Philippines, *kumri* in India, *cadang* in Malaysia and Indonesia, *lua* in Vietnam, *parcelero* in Puerto Rico, and *shamba* in Kenya. Under unregulated conditions shifting cultivation results in the destruction of extensive tracts of forest but if it is controlled and used for the formation of plantations of valuable trees, this otherwise harmful practice can be turned to very good account. The so-called *taungya* plantations of Burma, begun in 1869 and now covering a large area, were created with the aid of temporary cultivation, usually of hill rice, on forest clearings in which all unmarketable timber was felled or burnt; in the process the soil was enriched by the ashes which contained substantial quantities of the major elements, including nitrogen (Sanchez 1976) and minor elements. The plantations were chiefly of *Tectona grandis*, which was introduced along with the field crop. These plantations were formed efficiently and cheaply, while the cultivators provided local forest labour and grew their own food, a matter of great importance in tracts where communications are not good.

After its introduction to Burma by European foresters, the systematic regeneration of clear felled areas in conjunction with a temporary agricultural crop spread to India, eastern, central and west Africa, tropical America, and elsewhere in Asia because it can play a part in controlling shifting cultivation. The word *taungya* means 'hill cultivation' and was originally the Burmese word for shifting cultivation.

Clear cutting with artificial regeneration by *taungya* cultivators, supervised by staff of the forest service, is done in the tropical semi-evergreen and sal (*Shorea robusta*) forests of Chittagong in Bangladesh. The object is to replace the existing mixed forests with more or less uniform crops of *Tectona grandis, Lagerstroemia speciosa*, and *S. robusta*. One-third to one-fifth of the total forest area is selected for felling during a period of 20 years. The merchantable produce is removed from each annual coupe of 12 to 20 ha and the slash is burned. Strips of the existing forest are retained between the new crops as protection against fire and the build up of injurious insects (Directorate of Forests 1974). The trees and field crops are planted or direct seeded and during the first year the weeding of the agricultural crop also benefits the young trees. Where field crops are not raised subsequently the trees must be released from the competition of weeds thrice during the second year, twice in the third year and once in the fourth year after planting. In some parts of Bangladesh a cover crop of the legume *Tephrosia candida* is sown between the rows of *Shorea robusta* seedlings when these are 1 year old. During the second and third years the legume must be cut back and thinned so that it does not suppress the *Shorea*; in the fourth year it is removed. In later years climbers are cut when the *Shorea* crop is cleaned and thinned.

Taungya cultivation was introduced to southern Nigeria in 1927 and by 1976 24 427 *taungya* cultivators grew crops on almost 20 000 ha of forest reserves; in addition almost 1220 workers employed by the Forest Departments cultivated a further 1448 ha (FAO 1979a). The tree crop was usually teak for timber or *Gmelina arborea* on a rotation of 8 years for pulpwood. The main food crops grown during the 2 years with the trees were maize (*Zea mays*), yams (*Dioscorea* sp.), cassava (*Manihot esculenta*), and vegetables. Between one-half and two-thirds of the field crop went to feed the cultivators and their families. Access to fertile soil was very important and the demand for more land for *taungya* was high, although most of the cultivators also had to add to their income by growing cacao, kola, or rubber on other land or by taking paid work.

Destruction of forests by shifting cultivation is a serious problem in Thailand, particularly in the northern and north-eastern regions. The object of the forest village scheme, begun in 1968, is to encourage landless people to settle in communities which can offer them a better standard of living and greater stability than nomadic life. Each forest village is intended to include about 100 families and each family is allotted 1.6 ha of forests each year to be cleared and cultivated under *taungya* for 3 years. Progress with the forest village scheme was slow at first but by 1973 almost 2000 ha .of plantations were being established by *taungya* cultivators each year. In 1976 there were 21 forest villages with 817 families and 10 000 hectares were planted with trees (Samapuddhi 1974; FAO 1979a).

The administration of *taungya* in reserved forests requires firmness and discipline on the part of the forest services, otherwise it falls into disrepute. If the farmers and foresters understand and trust each other its potential for increasing the production of food, fuel, and timber is high and its contribution to the solution of local social problems can be considerable. The proximity of plantations to villages and roads facilitates subsequent tending and harvesting of fuelwood and timber. Indirectly *taungya* protects the natural indigenous forests, and in many parts of the tropics the formation of tree crops by *taungya* cultivators also provides foresters with managerial and silvicultural experience and data on yields of produce, which are valuable in planning and executing large-scale afforestation programmes.

Clear cutting with natural regeneration

Under certain conditions clear cutting may be followed by successful natural regeneration (see page 92). This may be obtained either from seed already on the area, or from seed disseminated by trees outside and usually adjoining it.

Regeneration from seed already on the area

The best known case of natural regeneration following clear cutting is in the Maritime pine forests of the coastal dunes of the Landes in south-western France. Clear cutting is done over large coupes of 80 to 100 ha and natural regeneration springs up in abundance from seed, some of which is already on the ground, but much of which is released from the cones of felled trees in April or May. Natural regeneration is especially favoured by the abundant annual seeding of the pine, the loose sandy soil, and the complete exposure to light of a species that demands it. After felling and harvesting a coupe the branchwood is spread evenly over it so as to distribute the seed; this tends to produce an even stocking of the area with young plants. After 4 years a weeding-and-cleaning is done leaving the young pines spaced at about one metre square, or 10 000 plants per hectare (Lanier 1986).

The inland or mountain form of Lodgepole pine provides another example of regeneration from seed already on the area. Lodgepole pine produces seed prolifically and good crops occur at intervals of 1 to 3 years. Throughout most of the inland distribution the cones are generally serotinous, although many exceptions are found. In the Rocky Mountain and Inter-mountain regions the closed-cone habit is widespread but in the Sierra Nevada, Oregon Cascades, eastern Siskiyou mountains, and in southern and Baja California trees are found with cones that open at maturity. Where the cones are serotinous the main supply of seed for natural regeneration comes from cones attached to the slash or knocked from it and scattered over the ground. Most of the seed is released after the cones have been exposed for 1 year, but further release continues for 6 or more years. Fire is not a prerequisite for the release of seed from cones (Bates *et al.* 1929; Tackle 1954). Usually the resin that seals the scales together melts at 45°C allowing them to flex and spread apart; this can occur without fire when enough heat reaches the surface of the cone through radiation, convection, or conduction (Crossley 1956). However, fire may hasten the opening of some cones that are not suitably positioned for the resin bond to be softened by solar heat. Best germination of the seed occurs in full sunlight on mineral soil or disturbed humus free of competing vegetation (Fowells 1965).

One of the most remarkable instances of natural regeneration on clear cut areas from seed stored in the ground was described by Troup (1952, p. 12) for *Tectona grandis* in Burma and southern India. The seed is enclosed in a hard-shelled nut and retains its viability for many years. Where seed bearers are plentiful, the nuts accumulate in the ground but germination of the seed does not take place until the nuts are alternately heated by the sun and moistened by rain, when dehiscence takes place. This condition is secured by clear cutting which is followed by the appearance of large numbers of seedlings springing from seed which may have lain dormant in the ground for many years. However, survival of these seedlings can only be secured by repeated and costly weeding, so it is generally found less expensive to ignore the natural regeneration and plant up the coupe in regular lines so that weeding is made easier.

Regeneration from seed disseminated from outside

This form of regeneration applies chiefly to species with light or winged seeds that are disseminated by wind, as in the case of many pines and other conifers as well as broad-leaved species like Red alder, Paper birch, and invasive tropical species which habitually spring up in quantity on recent clearings (Whitmore 1984). Less commonly, water or animals may be the agents of dispersal, as in the case of the irrigated plantations of the Punjab in Pakistan where mulberry sprang up extensively on cleared areas from seed spread by irrigation water and by red-breasted starlings *Sturnus roseus* (Khattak 1976).

In the Pacific coast region of North America, Douglas fir and its associates Western hemlock, Western red cedar, and Grand fir often regenerate profusely after clear cutting, particularly if the slash is consumed by fire immediately after felling and before any dormant seed in the ground has time to germinate. It is believed that under natural conditions the Douglas fir stands were regenerated after lightning fires (Smith 1986), the seed coming from large scattered trees which survived the fires on or around the burnt areas. In the absence of fire, old even-aged stands of Douglas fir tend to disintegrate very slowly and are replaced by the more shade-tolerant Western hemlock, Western red cedar, and Grand fir (Munger 1940); Douglas fir does not regenerate under these conditions because of the heavy shade (Williamson and Twombly 1983).

Patch felling

Fr. Exploitation par blocs.

This is a modification of the clear cutting system developed in the Pacific coast region of North America whereby patches of 16 to 80 ha are clear cut as single 'settings' of the machines used in logging, with the object of ensuring that a high proportion of Douglas fir appears in the new stands. The size of patch felling used for natural regeneration varies from 16 to 24 ha; early restocking of larger patches almost always depends on artificial regeneration (Smith 1986). The patches are separated for as long as possible by areas of living forest so as to secure good dispersal of seed and avoid the serious hazard created by large continuous areas of slash, particularly as regards fire and insect pests. The arrangement of the patches must also be integrated into the network of roads used to transport logs to the mills; extension of roads to small widely scattered patches is unsatisfactory and expensive.

Where the high lead system of skidding is used, most of the defective and otherwise unmerchantable trees are also felled to clear the patch. Controlled broadcast burning is done in weather that does not favour hot fires, to reduce the large amounts of slash, eliminate advance growth of the shade-tolerant species, and reduce the thickness of the layer of humus. Given a good supply of seed, satisfactory restocking of Douglas fir generally appears on shaded northern aspects, but on southern aspects the broadcast burn can cause excessive exposure of the mineral soil and create micro-climatic conditions that are too extreme for adequate natural regeneration of Douglas fir. As already noted on page 78, some shade is essential on southern exposures to protect newly germinated seedlings from injury by heat. Once established, Douglas fir seedlings grow best in full light; where Red alder has invaded the site it must be controlled because of its very rapid early growth.

Strip-like clear cuttings

Fr. Coupes par bandes; Ger. Streifenschlag: Span. Cortas por fajas

In Europe clear cutting with natural regeneration from adjoining stands or belts has been systematized chiefly in the case of Scots pine, although it has been applied to other conifers, including Austrian pine. The coupes are in the form of strip-like clear cuttings made either by progressive fellings or by fellings in alternate strips. It is a question whether a strip-like clear cutting should be classed as a form of the clear cutting system or as a form of the strip system. It is most logical to reckon as a strip felling one which is so narrow that the adjoining old stand exercises a powerful influence on the strip in regard to soil moisture and other physical factors which affect seed germination and the growth of seedlings; the width of such a strip would be about half the height of the adjoining old stand or not much more. A strip-like clear cutting, on the other hand, is not so influenced except possibly along the immediate edge of the adjoining old stand. It follows, therefore, that strip fellings in the strict sense apply generally to species like Norway spruce which are sensitive to drought in early youth and may benefit from lateral protection from the sun, rather than a hardy species such as Scots pine, the seedlings of which demand much light and may be adversely affected by close proximity to an adjoining stand.

Progressive fellings

Fr. Coupes progressives en bandes; Ger. Kahlstreifenschlag; Span. Cortas por fajas progresivos

Under this method, which is a very old one and still used in central Europe, the fellings proceed in a definite direction in the manner already described for the clear cutting system (page 66). If the wind direction at the time the seed ripens is at all constant, it is a great advantage to arrange the fellings so as to proceed against this direction; by such an arrangement seed is blown on to the adjacent cleared strip, the breadth of which should be such that seed in abundance will reach every part of it. The actual width of the strip must be determined by local observation, since even in the case of one and the same species the distance to which the seed is disseminated in sufficient quantity must depend largely on the speed and direction of the wind under average conditions. A mathematical model was devised by Ek *et al.* (1976) to determine width of strip for Black spruce in Canada.

The interval between successive adjacent fellings will depend on the frequency of seed years and the readiness with which regeneration appears and establishes itself. If one strip is cleared, the one adjoining it should not be cleared until the original one is fully regenerated, and a liberal margin of time should be allowed for this. Assuming, for instance, that under the prevailing seeding conditions a period of 3 or 4 years is found to be sufficient to ensure the establishment of natural regeneration over a cleared strip of a certain width, then it should be safe to allow a period of, say, 6 years between the cutting of one strip and the cutting of the one adjoining it. In an example given by Troup (1952, p. 14) the area to be regenerated is divided into six cutting sections with three coupes of the required width in each, so that it will take a period of 18 years to fell and regenerate the whole. In the last coupe of each cutting section a belt of seed bearers should be left along the windward side to regenerate the coupe, after which the belt should be felled and the blank strip regenerated artificially.

More flexibility can be introduced into the working by allocating a number of areas for regeneration during a given period and by not rigidly fixing the interval between successive fellings; this allows for the distribution of fellings over several areas, which will tend to equalize the area felled each year without too close an adherence to a fixed scheme, to which the vagaries of natural regeneration may not always lend themselves.

Fellings in alternate strips

Fr. Coupes par bandes alternes; Ger. Kulissenhiebe; Span. Aclareos sucesivos por fajas alternos

A method which has been tried from time to time is that of felling parallel clear strips, usually about 40 to 60 m wide, through the crop to be regenerated leaving intervening strips of smaller width untouched. The cleared strips regenerate naturally, and the young crop is allowed to grow to an age when it is capable of producing seed in quantity, when the intervening old strips are clear felled and the process is repeated. This method has not proved satisfactory. Apart from the difficulty of controlling it owing to the presence of two widely different age classes distributed over the area, there is considerable waste of space, particularly in the case of pines, owing to the failure of regeneration along the edge of the strips left uncut. Glew (1963) looked at the application of the method to White spruce in the northern interior region of British Columbia. He concluded that if the uncut strips were to be left for, say, 40 years there would be serious losses from windthrow, disease, and insect attack.

A more workable adaptation of this method is to make clear cuttings in parallel clear strips 40 to 50 m wide, belts of untouched forest 15 to 20 m wide being left between them to furnish seed (Fig. 10). Natural regeneration establishes itself on the cleared strips in full overhead light and when the young crop is sufficiently advanced, i.e. about 6 years after the first felling, the intervening belts of seed bearers are felled and the blank strips are regenerated artificially by direct seeding or planting. Because seed is distributed on two sides little or no account need be taken of wind direction in cutting the strips, but they should be aligned more or less at right angles to the roads so that when the seed bearers are felled, the timber does not have to be extracted through the young crop.

Figure 10.   Clear cutting by alternate strips with natural regeneration. Felled strips 30 m wide; standing strips 15 m wide. Scots pine, Forest of Bord, France.

Advantages and disadvantages of the clear cutting system

The advantages are:

1. It is the simplest of all high forest systems and readily lends itself to innovation in silviculture and forest management (see page 54).
2. It allows amelioration of the deficiencies of a site by cultivation, drainage, supplementary nutrition, and other means.
3. It allows improvement of the yield and quality of the new crop through introduction of superior indigenous and exotic species, provenances, and cultivars.
4. It affords complete overhead light, an important consideration for light-demanding species.
5. As a rule, a well-stocked crop is established more readily and rapidly under the clear cutting system than under systems in which regeneration is established by degrees.
6. Owing to the even-aged condition of the crops with assured stocking density the clear cutting system can produce more stems with low rate of taper and smaller branches than are produced by the more uneven-aged systems.
7. Felling and extraction of the timber is completed before the new crop regenerates, so no damage is done to it.
8. It represents the utmost concentration of work, every tree on the coupe being felled; this means a larger out-turn per hectare and therefore greater economy in harvesting, than in systems where only a portion of the trees are felled at one time.

The disadvantages are:

1. On steep hillsides it may expose the site to erosion and where the soil is unstable to landslips, while it affords no safeguard against the rapid run-off of precipitation.
2. The complete clearance of the forest cover may produce conditions of microclimate, soil, competing weeds, and predators adverse to the survival and growth of young plants of the desired species.
3. It results in accumulation of large quantities of slash after fellings. Where utilization is not intensive and tops and branches of some size are left lying on the site, this slash forms the breeding ground of weevils and beetles, which are destructive more particularly of conifer forests. This risk can be avoided by more intensive utilization and by burning or macerating the slash.

4. It prevents full use of the potential for growth and yield of timber of individual trees, which is a feature of uneven-aged systems. This may be overcome by the use of clonal cultivars.
5. It possesses the disadvantages of even-aged systems in producing a type of forest less resistant to damage by snow and wind than the more uneven-aged systems.
6. From the aesthetic point of view, careful design of the forest is required to reduce the impact of the clear cutting system on the landscape.

The disadvantages just mentioned are intensified if the coupes are large, or the site remains uncovered for some time before the young crop closes up; they are reduced to a minimum when small or carefully shaped coupes are adopted, and where the site is speedily reclothed with trees.

Application in practice

The progressive clear cutting system with natural regeneration from adjoining stands appears to have been in operation in Germany for some hundreds of years, the coupes being long and narrow for seeding purposes. In Europe the clear cutting system began to receive serious attention about the middle of the eighteenth century; its effectiveness and the sureness of its results brought it more and more into favour, so that by the end of the century it was in full swing. It was in Germany, however, and in countries influenced by German ideas, that clear cutting came to be most widely practised. Oak was among the first species employed to re-afforest clear cut areas and many fine oak stands were raised by dibbling acorns, often in conjunction with field crops. During the latter half of the eighteenth century and subsequently, much re-afforestation of sandy soils was done by direct seeding of Scots pine, but it was Norway spruce that became the species most widely associated with the clear cutting system.

Heinrich von Cotta (born 1763, died 1844) introduced the clear cutting system in the state of Saxony at the beginning of the nineteenth century. He saw it as the only means of regenerating forests which had suffered to such an extent from unregulated exploitation fellings, excessive grazing, and removal of litter that natural regeneration was out of the question. His major innovation was to form cutting sections with the object of providing protection against wind. According to Mantel (1964) Cotta's publications, including 'Anweisung zum waldbau' first published in 1817, contained 'the essence of the young science of forestry in the first half of the nineteenth century'.

Today clear cutting in its various forms is the most widely used silvicultural system in the world. It is simple, and can be adapted to many conditions of climate, terrain, soil, and tree species. It also lends itself readily to technical innovation ranging from the use of improved cultivars to new harvesting systems and even to agro-forestry (see page 242). These great advantages make it certain that clear cutting will continue to be widely used. Like all simple and effective methods it has been applied where conditions are unsuitable, with poor results; there have also been shortcomings in the techniques employed. In many countries, however, clear cutting is being applied on a sound ecological and economic basis. Teams of specialists representing geology, hydrology, civil engineering, wildlife management, fresh water fisheries, and landscape architecture are working with foresters to plan and design forests worked under the clear cutting system.

Although typically an even-aged system, clear cutting does not demand the growing of pure crops, as is sometimes alleged against it. It is true that pure crops are produced in some of the best known examples of clear cutting with natural regeneration, but in the case of Scots, Austrian, Maritime, and Jack pine, to quote only four instances, a shelterwood system in the same localities would also produce pure crops. When artificial regeneration is used, any mixture can be adopted that the site will permit and the circumstances demand; the chief difficulties are correct choice of species and the best method of forming the mixture.

A fine example of even-aged, mixed stands produced by the variant of the clear cutting system called patch felling (page 83) can be seen on land managed by MacMillan Bloedel Limited to the east of Port Albirni on Vancouver Island in British Columbia. Here a near-normal series of age classes is gradually being created out of indigenous forest. Each stand consists largely of Douglas fir with its associate Western hemlock, together with Western red cedar, Alaska yellow cedar, some Lodgepole pine, Red alder, and Bigleaf maple. A few miles to the east a reserve of natural indigenous forest can be seen in the MacMillan Park. The latter is a wonderful natural monument; the former is a well-managed and renewable natural resource.

7

Shelterwood systems

Fr. Coupe d'abri; Ger. Schirmschlag; Span. Corta de abrigo

Introduction

Shelterwood systems are those high forest systems in which the young crop is established under the overhead or side shelter of the old one; at the same time the old crop protects the site. The term 'shelterwood systems' includes systems of successive regeneration fellings together with the selection system.

In general though not universally, the shelterwood systems aim at natural regeneration whereas the clear cutting system, usually though not invariably, must rely on artificial regeneration. Under the clear cutting system, using natural regeneration, the area to be regenerated is cleared in a single felling and restocked in one season. However, this is possible only where conditions are unusually favourable for natural regeneration, and where it is not necessary to retain any part of the old crop to protect the young one against frost, drought, or other dangers. In the great majority of cases complete natural regeneration does not follow from a single felling, and protective cover for the seedlings of shade bearers may be needed for some years after they appear. In either case, part of the old crop has to be retained for a time to provide seed or protection. Thus, the old crop is removed by two or more successive fellings, and the new crop establishes itself from seed shed by trees retained on the area; these are called 'seed bearers'.

In some cases the new crop may be introduced artificially under the protection of the old one, the latter being removed subsequently in one or more fellings. The term 'systems of successive regeneration fellings' is used for those in which definite areas are regenerated by two or more successive fellings, known as 'regeneration fellings', extending over a period of years.

The distinction between the different systems of successive regeneration fellings lies not in their general framework but in the way the fellings are done and are distributed in time and space. The difficulty of devising a satisfactory classification of silvicultural systems applies with special emphasis to the systems of successive regeneration fellings, which show many variants, often merge into each other, and are frequently used in combination. European and North American terminology tends to confuse rather than assist, since the same term is often used by different writers in a different sense, while undue importance is sometimes attached to small variations. Following the classification given on page 5, the systems will be described in the following order:

the uniform system;
the group system;
the irregular shelterwood system;
the strip systems;
the wedge system; and
the tropical shelterwood system.

We will describe the uniform system in detail as the techniques used are based on principles that apply to all the others. Before proceeding there are two topics requiring some discussion: these are the advantages and disadvantages of shelterwood compared with clear cutting, and the use of natural regeneration in modern forestry.

The advantages of shelterwood systems are:

1. They provide protection to species sensitive in youth to frost, drought, and cold winds. Such protection is not provided under the clear cutting system except with narrow or small coupes.
2. The soil is more effectively protected than under the clear cutting system, particularly when the canopy is opened gradually and cautiously. There is less risk of desiccation and invasion by competitive weeds.
3. There is also less risk of multiplication of injurious insects that breed in clearings.
4. On steep and unstable slopes there is less risk of erosion or rapid run-off than under the clear cutting system.
5. By adopting certain forms of shelterwood system, more effective measures can be taken against damage by wind and snow than under the clear cutting system.
6. An opportunity is given to the best trees to put on added increment when opened out in regeneration fellings.
7. From an aesthetic point of view shelterwood systems are usually preferable to the clear cutting system, especially if the old crop is removed gradually.

The disadvantages of shelterwood systems are:

1. They require more skill and occupy more time than the clear cutting system.
2. Work is less concentrated and felling and extraction cannot be done so economically.
3. Damage is done to a greater or lesser extent by felling trees over young growth and extracting timber through it. The amount of damage varies with different systems and can almost entirely be prevented by suitable methods and thorough planning.
4. In some cases the young crop takes more time to establish itself than under the clear cutting system. This loss of time may incur serious economic loss.
5. The rate of cutting and regeneration are more difficult to control than under the clear cutting system.

The use of natural regeneration

As we noted when discussing forest ecology (page 11), advocates of the 'natural' approach to forestry require that silviculture should be based on the ecology of natural indigenous forests, so it is appropriate to begin this discussion of natural regeneration by summarizing the advantages seen from this point of view:

1. Regeneration takes place under the protection of the old crop and this more or less approaches the process in natural indigenous forests.
2. The microclimate at and near the ground is likely to be favourable to the needs of seedlings, at least during the early stages of their development.
3. The layer of humus covering the upper horizons of the soil provides a good medium for germination protected from drying wind and excess sunlight, and favourable to the early survival of seedlings.
4. The seed bearers used as parents of the succeeding crop are well-adapted to the site. This is the most frequently cited advantage and it is usually, but not invariably valid. Well-adapted populations are not always the most productive nor the most valuable in their silvicultural characters or timber qualities (see page 21).
5. Mixtures of species can more readily be obtained and be closely matched to local site variations.
6. The step-wise progress to a multi-storied structure is generally easier. This is important when irregular stands are desired.

7. The interruption in production associated with clear cutting is shortened or absent. Considerable increment in value as well as volume often accumulates on the stems of the seed bearers during the course of regeneration fellings.

The disadvantages of natural regeneration are mainly managerial and economic:

1. It is rarely an easy operation and is always expensive in skilled man-power, time, and money.
2. It is rare not to have to aid the process of natural regeneration so as to reduce the risk of failure, correct deficiencies in stocking, or shorten the duration of individual operations.

Deliberate combination of natural and artificial regeneration is an old and well-established practice. The present trend is toward assisting natural regeneration by direct seeding or planting whenever this is required. Similarly many foresters who are committed to artificial methods will accept suitable natural seedlings when they appear.

In degraded or mismanaged forest containing few seed bearers and many weed-covered blanks natural regeneration usually has to be ruled out. Clear cutting with artificial regeneration will lead directly to a healthy and pro-ductive growing stock (see page 234). In nature reserves and other areas of special interest natural regeneration may be desired so that particular ecosystems may be conserved. If it fails or proves very difficult, recourse may be had to direct seeding or planting using seed derived from the existing population on the site so that the indigenous provenance or race may be perpetuated.

The process of natural regeneration consists of a series of interrelated events, each greatly influenced by chance. Much can be done, by actions founded on observation, to assist the process and reduce the impact of factors threatening failure so that it can be brought to a successful con-clusion in a reasonable time, but natural regeneration remains a precarious stage in the life of a forest stand (Lanier 1986, p. 146). The requirements for successful natural regeneration are:

a regular supply of viable seed;
a receptive seed-bed well-supplied with water and nutrients;
a microclimate favourable to germination of seed and survival and growth of young plants;
resistance of plants to competing vegetation and to damage by animals, insects, fungi, and extremes of climate.

These requirements are very similar to those listed on page 76 for direct seeding. The differences lie in the use for natural regeneration of mature trees and crops on or near the area to be regenerated to provide seed and modify the microclimate at or near the forest floor.

Much detailed information about flowering and seed production has been gathered by tree breeders in their research into the factors affecting flower bud initiation and into methods of reducing the losses that occur between flowering and ripening of viable fruits and seeds. The over-riding influence of climate on periodicity of flowering and seed production in the temperate regions has been confirmed (Matthews 1963, 1964; Sweet 1975). Favourable sites for most tree species have dry, warm summers and soils that are slightly below the optimum for the vegetative growth of a given species. There is also a strong genetic component in flowering and seed production; trees chosen as seed bearers should show evidence of having produced cones, fruits, and viable seed on previous occasions. The crowns of seed bearers must be well-exposed to sunlight. Applications of fertilizer have increased flowering and fruit production. The most important element is nitrogen and the precise timing of the application must be determined for each combination of species and site; it appears to lie between late June and late August. On dry sites irrigation can increase flowering and fruiting but timing is particularly important for this.

Where conditions of seeding and germination are especially favourable, natural regeneration over large coupes can be done with success. Where seeding is sporadic or irregular natural regeneration over smaller coupes is indicated, and the selection or irregular shelterwood systems often give better results. An example of favourable conditions is provided by Yellow poplar, a species of the eastern United States of America occurring most abundantly in the valley of the Ohio river and on mountain slopes in North Carolina, Tennessee, Kentucky, and West Virginia. Because of its seed production, Yellow poplar is readily regenerated by natural means, as the following quotation from Beck and Sims (1983) makes clear:

'The first requirement, adequate seed, is seldom a problem on sites capable of growing Yellow poplar even where only a few seed-producing trees are available. Large seed crops ... are produced annually in Yellow poplar stands. Viable seed is disseminated from mid-October to mid-March; the viability, which ranges from 5 to 20 per cent, is about equal throughout the seed fall period. The individual winged samaras may be scattered by wind to distances equal to four or five times the height of the seed bearer. Distribution of seed is satisfactory up to 60 m from a good tree in the direction of the prevailing wind and 30 m in all other directions. In addition, seed may be stored in the forest floor for up to 8 years and germinate and grow when proper conditions of light and moisture are created.'

In the Scandinavian countries considerable use is made of mechanical scarifiers to prepare seed-beds for natural regeneration and direct seeding (page 72). Hagner (1962) examined the occurrence of natural seedlings of Scots pine and Norway spruce regenerated under the uniform system on 57 sites between 62 and 64°N latitude in Sweden. He found that scarification was necessary to obtain satisfactory stocking of Scots pine seedlings on the better sites. On such sites re-invasion of weeds made patches greater than 1.2 m^2 essential because occlusion took only 5 years on smaller patches. Scarification was not so essential for Norway spruce if a good seed crop fell immediately before or soon after the seeding felling, but it did give some advantage.

Scarification or controlled burning which puts seed of Yellow poplar in contact with mineral soil, significantly increases the number of seedlings when compared with an undisturbed forest floor (Beck and Sims 1983). However, under normal conditions, the site disturbance caused by harvesting a mature stand is the only seed-bed preparation needed to provide enough Yellow poplar seedlings for a new stand. The amount of regeneration obtained increases directly as the intensity of the felling, with clear cutting giving the highest stocking.

An example of intensive seed-bed preparation comes from the Forêt de Soignes near Brussels in Belgium (Reade 1965). Here beech is regenerated naturally under the uniform system in mature stands 100 to 140 years old. A favourable ground covering includes Wood anemone *Anemone nemorosa*, Yellow archangel *Lamium galeobdolon*, Lesser celandine *Ranunculus ficaria*, Wood sage *Teucrium scorodonia*, and foxglove *Digitalis purpurea*. The cover of grasses should be neither strong nor dense. Unfavourable conditions are represented by blackberry *Rubus fruticosus*, raspberry *R. idaeus*, and bracken *Pteris aquilinum*. Dense woodrush *Luzula sylvatica*, Tufted hair grass *Deschampsia caespitosa*, and Soft rush *Juncus effusus*, are also unfavourable to natural regeneration of beech. Preparations for regeneration begin in late summer when a good mast is seen to be forming on the trees. The seed-bed is prepared by cultivation to depths of 6–8 cm using a tractor-drawn rotovator. This work must be completed by mid-October. As soon as the beech nuts have fallen the ground is harrowed to cover them with soil as protection against Wood pigeons (*Columba palumbus*).

Since the German silviculturist Wagner (1912, 1923) published the results of his studies on the effects of lateral shelter from the old stand on the microclimate in strip fellings, numerous similar ecological studies have been made to determine the best orientation and size of strip and group fellings for natural regeneration of a given species on a particular site. Wagner gathered his data and developed his ideas at Gaildorf in Baden-Württemburg, West Germany whilst managing mixed forests of Norway spruce, Scots pine, and beech. The natural regeneration of Norway spruce gave most trouble. Although the seed germinated well the seedlings, owing to their shallow root systems, died of drought during the dry weather usually prevalent in June. Wagner found that agencies bringing moisture to the soil (rain, snow, and dew) were generally favourable to the regeneration of Norway spruce while those that had a desiccating effect on the soil (sun, wind, and frost) were unfavourable. His analysis of the microclimate on the different sides of mature stands revealed that the north-west and north sides, and particularly the former, were most favourable to regeneration, the north-east side coming next. Other aspects were unfavourable, from east through to south being decidedly sterile. Wagner therefore decided that the ideal direction in which fellings should proceed at Gaildorf was from north-west to south-east, but where there was danger from wind it was necessary to alter the direction to one of north to south (see page 138).

Turning now to the survival and growth of young regeneration, the germination of Yellow poplar seed is followed by several critical years for the seedlings (Beck and Sims 1983). Although considered to be intolerant of shade, Yellow poplar seedlings reach maximum or near-maximum photosynthetic efficiency at relatively low light intensities. During the establishment period protection from drying, frost-lift and flooding may be more important than light requirements. After the first few growing seasons, competition from vegetation can become the most important factor affecting survival and growth of young regeneration.

Competition for light, water, and nutrients can come from the seed bearers. Hagner (1962), who was quoted earlier in relation to scarification, found that a uniform shelterwood did protect Norway spruce seedlings from frost damage and frost-lift but the height increment of the regeneration was inhibited by the seed bearers. In regions with hot, dry summers seedlings of many pine species clearly suffer from the competition for water and nutrients exerted by the seed bearers through their roots; thus the silvicultural prescription recommends speedy removal of the seed bearers when the regeneration is established (see page 124).

The problem of damage to young regeneration by animals can be approached by controlling the populations of the animals concerned, excluding them by fencing from areas being regenerated, and protecting a certain number of individual plants by erecting shelters of suitable design. Edwards (1981) followed the progress of several hundred seedlings of Scots pine regenerated under a uniform shelterwood of 100 mature seed bearers.

His observations were made on a site at 300 m elevation on a moderate, south-facing slope in the indigenous Scots pine forest of Glen Tanar, near Aboyne in north-east Scotland.

Removing the surface vegetation to expose the layers of humus and produce a tilth improved the temperature and moisture regime of the seed-bed and stimulated germination. Seed-eating birds and rodents caused considerable losses of seed before germination and the numbers of newly germinating seedlings were severely depleted during the first month due to grazing by slugs (*Gastropoda: Pulmonata*) and desiccation, but mortality declined as the stems of seedlings became lignified and their taproots penetrated moist humus and mineral soil. There was evidence that complete exclusion of Red and Roe deer in the early stages of regeneration was not necessarily an advantage because they grazed the ground vegetation, but as the plants grew taller, browsing by the deer and capercailzie (*Tetrao urogallus*) could keep them to the level of the tallest vegetation, that is, at about 30 cm, and prevent further growth. Because Red and Roe deer and capercailzie are a sporting asset on this private forest, fencing of the area being regenerated would not be done unless this restriction of height growth occurred.

Where regeneration of the principal species is sparse, individual shelters are a possible means of protection provided they are cheap, easy to erect, and offer other advantages besides protection from animals. The kind of tree shelter now being used in Britain in very large numbers is a tube of transparent or translucent plastic, usually 1.2 m tall and 8–10 cm in cross section, which surrounds each tree for about 5 years (Evans 1987). Tree shelters create a favourable microclimate around the tree which greatly increases height growth of many broad-leaved species. They protect the tree sufficiently to permit weeding with chemical herbicides without great risk of damage to it. The shelter is left around the tree until it disintegrates; a fully biodegradable plastic is being sought.

The great diversity of species and ecosystems in tropical moist forests and the lack of detailed knowledge about the silvicultural characters of many principal species and those associated with them means that natural regeneration of the principal species can rarely be induced. At present foresters in the tropics must rely on the presence of advance growth to regenerate indigenous forests (Wyatt-Smith 1987). Whitmore (1984) has summarized much of the existing knowledge about seed production, seed dispersal, and seedling establishment in tropical moist forests of the Far East, and if these and similar forests elsewhere are to be regenerated by natural means such research must continue.

8

The uniform system

Fr. Méthode de régénération des coupes progressives; Ger. Schirmschlag-betrieb; Span. Methodo por aclareos sucesivos

The term uniform system is an abbreviation of 'shelterwood uniform system'. It implies a uniform opening of the canopy for regeneration purposes, as well as an even-aged and regular condition of the young crop produced (Robertson 1971). Another name is 'shelterwood compartment system', denoting that the canopy is opened out for regeneration over whole compartments at a time. In North America the term 'seed tree method' is used for a form of the uniform system in which seed trees are widely spaced over the whole area (Smith 1986).

General description

We will begin with a young even-aged crop of oak or beech and follow it through the stages of its life until it reaches maturity and is removed to make room for a new crop. In its early stages, from 3 to 10 m top height, the crop is cleaned, that is, unwanted species such as birch, aspen, and willow, as well as trees of the principal species with crooked stems and large branches are removed. Apart from this the crop is kept dense so as to kill off side branches and encourage the development of clean stems. In France the work of weeding and cleaning is done from a network of racks 3–4 m wide (which delimit blocks of 100 × 50 metres or 0.5 ha) and narrow tending paths 1 or 2 m wide and 10 to 12 m apart running at right angles to the racks (Oswald 1982).

Periodic thinnings to remove poorer stems and favour the better trees begin when the crop reaches top heights of 10 to 14 m for beech and 10 to 16 m for oak. Trees that will form the 'main crop' are chosen when the whole crop has reached top heights of 15 m for beech and 18 m for oak. The thinnings are intended to stimulate crown development so as to maintain diameter increment and encourage seed production. However, it is equally important that the canopy should not be broken permanently at this stage, as this is likely to result in strong growth of weeds and grass which may prevent natural regeneration when it is wanted.

As the crop approaches the age at which it will be felled and regenerated, it should consist of trees with long, straight stems free from branches and with well-developed crowns forming a complete canopy and capable of producing good crops of seed. Trees with good crowns will have correspondingly well-developed root systems, so that when the crop is opened out they should be reasonably wind-firm. Owing to the closed canopy, the ground cover should consist at most of a sparse growth of shade-enduring plants.

The time has now arrived for the regeneration fellings (Fig. 11). These consist of the seeding felling, which opens the canopy to provide sufficient light to ensure survival for a short time of seedlings springing up from seed shed by the mature trees overhead; and secondary fellings, in which the seed bearers are removed in one or more fellings at suitable intervals to admit more light to seedlings on the ground. The last of the secondary fellings is termed the final felling; this is made when the young crop is well established.

Figure 11.   Shelterwood uniform system, with good natural regeneration of beech following one seeding felling and one secondary felling. Now ready for the final felling. Lyons-la-forêt, France.

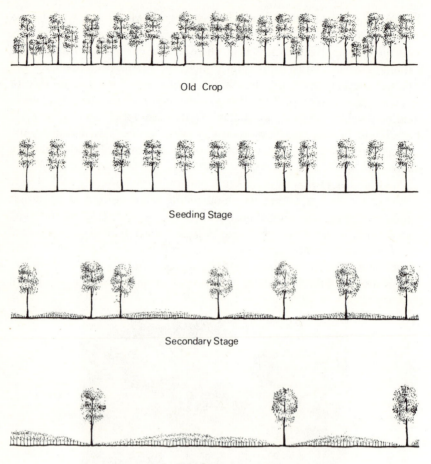

Old Crop

Seeding Stage

Secondary Stage

Final Stage

Figure 12.   Shelterwood uniform system showing successive stages of regeneration in beech.

Figure 12 shows diagrammatically the different stages in the process of regeneration. A crop in which the seeding felling has been done is said to be in the seeding stage; one in which secondary fellings have already begun is in the secondary stage; and one which awaits the final felling is in the final stage.

At one time it was customary to make a preparatory felling shortly before the seeding felling, or in certain cases two or more preparatory fellings extending over a period of up to 10 years. The object of these was to promote crown development and seed production, allow light and heat to reach the ground so as to hasten decomposition of the humus layer, and produce a good seed-bed by exposing mineral soil. Hartig (1791) recommended preparatory fellings in places where seed crops are rare, so that when a good mast year occurred a large area would be ready for seeding fellings. However, it is now recognized that crown development cannot be secured satisfactorily by a single felling, or even by two or more fellings, done a short time before the seeding felling. The correct time to promote crown development is throughout the greater part of the life of the crop from the small pole stage onwards. In France, the *coupe préparatoire* is now understood to mean the last thinning done before the seeding felling (Buffet 1980, 1981).

It follows from what has been said above, that the life of a crop treated under the uniform system—and this applies to shelterwood systems in general—is divided into two parts, namely:

(1) education or preparation (Lanier 1986) in which the crop is subjected to early cleanings followed by periodic thinnings extending over a large portion of its life and fitting it for regeneration; and

(2) regeneration, beginning with the seeding felling and ending when the final felling has been done and the young crop is fully established.

Great attention is paid to the 'education' of the crop so as to bring it to a condition that will favour regeneration when the time comes. Thinnings at intervals of 5 to 15 years are made throughout the education phase, these thinnings often increasing in intensity up to the time of regeneration so that the transition from education to regeneration may be almost imperceptible. In the case of shelterwood systems generally, at least half the total production of a crop should be yielded by thinnings.

Regeneration fellings

Fr. Coupes de régénération; Ger. Verjüngungshiebe; Span. Cortas de regeneración

The seeding felling

Fr. Coupe d'ensemencement; Ger. Besamungshieb, samenschlag, besamungschlag; Span. Corta de sementara

If thinnings have been done properly through the life of the crop, by the time regeneration is desired the trees are able to bear plentiful crops of seed and the soil is in a receptive condition, free from excessive weeds or a thick layer of undecomposed humus. The canopy will be closed and the trees themselves will be the best seed bearers available, the different species in a mixed crop being as far as possible in the desired proportion. The seed bearers should have reached the minimum but should not exceed the maximum age of abundant seed production.

The object of the seeding felling is not to stimulate seed production but allow long-wave radiation to warm the soil so promoting germination, and admit sufficient light to enable young seedlings to survive for 2 or 3 years, or until it is necessary to admit more light by further opening the canopy. With most species success is best assured by making seeding fellings only in good seed years, and this rule is commonly observed where:

good seed years occur at long or irregular intervals;
when the seed is of a species that loses viability quickly;
where opening the canopy is likely to stimulate strong weed growth (particularly a matted covering of grass) or cause deterioration of the surface layers of the soil.

In these cases it is generally desirable to await the ripening or even the fall of seed before making the seeding felling; this provides a sure supply of seeds and avoids the possibility of mistakes in the event of premature destruction of the seed crop by gales or other calamities. Good seed years need not necessarily be awaited:

where they occur at frequent and regular intervals;
when seed stored in the ground retains its viability for some years; and
if opening the canopy is not likely to stimulate strong weed growth or soil deterioration before the young crop can establish itself.

Under favourable conditions seeding fellings are often made in oak, beech, and Scots pine forests without regard to the occurrence of seed years.

The ideal of the uniform system is to obtain even-aged masses of regeneration over whole compartments or other areas of some size. In France the compartments are from 15 to 30 ha. Great importance is attached to a high well-distributed canopy which admits light to the ground evenly and not in large patches of direct sunlight, because this is important in securing evenly distributed regeneration. Trees with straggling crowns and low branching, together with diseased or defective stems, are felled wherever possible. If any low-branching trees have to be retained as seed bearers or for cover, the lower branches may be pruned off. Small trees and even shrubs forming an underwood should be removed, since these can prevent light, heat, and rain from reaching the ground and so adversely affect regeneration.

Where the uniform system is practised in its ideal form under favourable conditions, advance growth is generally removed at the time of the seeding felling. Solitary saplings or groups of saplings, even if they recover from the suppression they have endured so far, develop into branchy 'wolf' trees and interfere with the regularity of the young crop. An exception to this rule is where large gaps have been created by wind or other agencies and have become filled with masses of promising regeneration of sufficient extent to form self-contained crops. Again, in the case of mixtures of shade-bearing and light-demanding species, promising advance growth of the former may be left while the regeneration of the latter is being secured by the seeding felling. Where conditions are less favourable and advantage has to be taken of any advance growth that may appear, we reach the border between the uniform system and the group system (see Chapter 9).

The extent to which the canopy is opened depends on species, climate, and other factors. For sensitive shade-bearers requiring protection from frost, drought, or cold winds, and for species with heavy seeds, the opening should be slight to moderate; for hardy light-demanders with small and, particularly, winged seeds the seed bearers may be widely spread, the actual spacing being based on the distance to which seed is dispersed in quantity by a moderate breeze.

Examples of seeding fellings applied to different species are given on pages 114 to 120. Where opening the canopy is likely to be accompanied by strong weed growth or rapid drying or deterioration of the surface layers of the soil, the seeding felling must be done with caution. More cover should be retained on hot slopes than on cool slopes, especially where the soil is dry and shallow. Where there is matted grass the ground may be scarified or cultivated in a good seed year to allow seed to reach the mineral soil. Admission of pigs (*Sus* species) for a time before the seeding felling can have an excellent effect in cultivating the soil. In conifer stands removal of a covering of undecomposed needles is often an essential aid to natural regeneration. In some tropical and sub-tropical forests, burning the layer of litter benefits natural regeneration, and some species will not regenerate successfully otherwise (page 158). Under favourable conditions one seeding felling is sufficient, but if complete regeneration does not follow, it may be necessary to make a second when the next seed year occurs.

Secondary fellings

Fr. Coupes secondaires; Ger. Lichtungshiebe; Span. Cortas intermedios

The object of secondary fellings is to remove the overwood and uncover the young crop so as to provide it with more light, water, and nutrients. When the young crop has been fully established as a result of the seeding felling, the overhead cover should not be retained a day longer than is necessary. Under the most favourable conditions an even-aged young crop, resulting from a single seed year, should cover the whole regeneration area from end to end. Where this happens in the case of a hardy light-demander the whole of the overwood may be removed in a final felling, within a few years of the seeding felling. In the case of a sensitive shade-bearer, however, the young crop should be uncovered more gradually by means of two or more secondary fellings carried out at intervals so that it may receive protection from frost, drought, or other risks which the overwood provides; as the young crop develops it is gradually freed from protective cover. Many years may elapse before the last trees of the overwood are removed.

In practice, however, conditions are not always so favourable as those just described. The seeding felling may result in successful regeneration in some places and failure in others, in which case another seed year, or more than one, may have to be awaited before regeneration appears over the whole compartment. In such circumstances secondary fellings do not proceed with mathematical regularity; some parts of the compartment may still be in the seeding stage while others have reached varying degrees of the secondary stage and, in places, even the final stage may have been reached. Thus in making secondary fellings, the state of the regeneration is the chief guide to the selection of trees to be felled or retained. Where regeneration is sufficiently well-advanced it should be freed from overhead cover, otherwise in the case of sensitive species it may be necessary to retain the cover for a few years longer. The rule is—follow the regeneration. Hence if the species concerned is deciduous, marking secondary fellings should be done when leaves are present, so that young trees are readily visible.

The final felling

Fr. Coupe définitive; Ger. Endhieb, räumungshieb; Span. Corta final

In the final felling all remaining seed bearers are removed and the fully established young crop remains. After the final felling blanks are restocked artificially, as a rule by planting. With sensitive shade-bearers it may be necessary to plant up blanks early in the secondary stage so as to protect young plants.

The number of secondary fellings and the intervals between them vary greatly according to circumstances. With a hardy light-demander like Scots pine, if complete regeneration is secured in a single seed year, only two regeneration fellings are usually needed, seeding and final, the latter following 3 to 5 years after the former. With a frost-sensitive shade-bearer like beech, in which the young crop is uncovered gradually, or in cases where regeneration establishes itself with some difficulty, three or more secondary fellings, done at intervals over 15 to 20 years may be necessary (see page 115).

Silvicultural and management aspects of regeneration fellings

When the old crop is opened out in successive regeneration fellings, the remaining trees respond by enhanced diameter increment called 'light increment' (Köstler 1956, p. 327). Hence, in the seeding and secondary fellings care should be taken to remove defective trees wherever possible and leave those with straight, clean stems; increment produced by the latter is of greater value than that produced by the former. In this way the quality of the remaining overstorey becomes progressively better until the finest trees of the crop are generally obtained from the later secondary fellings.

Another important reason for careful selection of the best phenotypes as seed bearers is to increase the probability of the young crop containing a good proportion of the best genotypes that are available. The degree of improvement, or genetic gain, depends on the intensity of selection and the narrow-sense heritability of characters affecting the properties of the timber produced (page 19). Stem straightness, length and angle of branches, and spiral grain are moderately to highly heritable so genetic gains are likely to arise from careful choice of seed bearers.

It is sometimes customary to ignore the distinction between the different categories of regeneration fellings, and to regard these as nothing more than a continuation of the process of thinning after regeneration has begun to appear. This is logical where much flexibility is allowed in choosing the areas in which regeneration is done and the period during which it is to be established; but where definite areas are set aside for regeneration in a given period of years, recognition of separate stages in the progress of regeneration is a great convenience, while there are also marked differences in the methods of executing seeding and secondary fellings.

A factor that strongly affects the conduct of regeneration fellings is the annual cut or yield, which is usually, though not invariably, fixed by volume. Circumstances must determine each year whether the prescribed volume shall be obtained from seeding or secondary fellings or from both. When an area is first taken in hand for regeneration every effort should be made to place as much of it as possible into the seeding stage, and hence at this time the whole yield will generally be obtained from seeding fellings alone; where seed years occur at quite long intervals it may even be advisable to exceed the prescribed yield during a good seed year. As regeneration proceeds the yield will be furnished mainly by secondary fellings.

In some cases the need to adhere to the prescribed yield may conflict with correct silviculture. If regeneration is backward, silviculture may require retention of the overstorey for seeding purposes or to protect the young crop, while management requires that fellings should be done to provide the prescribed yield. Again, regeneration may have established itself in great abundance, and may be in urgent need of release by removal of the overstorey; but if this is done the out-turn may considerably exceed the prescribed yield for a year or two, to be followed by a corresponding drop in out-turn during the subsequent years. At times it is not easy to hold the balance between the respective needs of silviculture and management.

Protective measures

Careless felling and extraction during the secondary and final stages may do much harm to young growth. The chief precautions taken to minimize damage are:

(1) fell and extract during the winter when there is deep snow on the ground;
(2) lop main branches or whole crowns of trees before felling;
(3) make extraction racks at intervals through the young growth, and use the same routes to extract a succession of logs.

In broad-leaved forest it is possible after the final felling to cut back damaged saplings to stimulate regrowth of straight shoots.

In hilly country a compartment or sequence of compartments is often worked over so that regeneration commences at the top of a slope and proceeds downhill to the extraction route in the valley. Despite the considerable damage that may be done to the young crop by felling and extracting timber, the young crop closes up and the results of damage disappear in a remarkably short time.

As a precaution against wind damage, fellings sometimes begin in the east of the regeneration area and proceed toward the west. In parts of the Black forest region of West Germany, this procedure was introduced in the early part of the nineteenth century; it developed afterwards into the shelterwood strip system (Chapter 11).

Periods and periodic blocks

Period; Fr. Période; Ger. Periode; Span. Periódo de reproducción
Periodic block; Fr. Affectation; Ger. Periodenfläche; Span. Cuartel especial

Under the uniform system it requires more than 1 year to regenerate a given area. In order to systematize operations and ensure that the whole forest shall be felled and regenerated during the course of a rotation, the plan commonly followed under favourable conditions is to divide the rotation into a number of periodic blocks, each to be felled and regenerated in turn during the course of successive periods. For example, if the rotation is 100 years, divided into five periods of 20 years, and if one whole rotation under the uniform system has just been completed, the normal distribution of age classes should be that shown in Table 1.

TABLE 1.   *Normal distribution of age classes*

	Age of crop in years		
Period and periodic block	At beginning of period (present time)	At end of period (20 years hence)	Notes
I	81–100	1–20	Old crop all removed; only young regenerated crop remains.
II	61–80	81–100	To be regenerated during period II.
III	41–60	61–80	
IV	21–40	41–60	
V	1–20	21–40	

Length of period

The period is the estimated number of years required to obtain complete regeneration over a whole periodic block and establish this regeneration up to the time when the overwood is no longer necessary. It should begin with the seeding felling and end when the last remaining seed bearers are removed in final fellings. It is necessary to distinguish between the actual time taken to regenerate small sections of forest and the absolute length of time usually taken to regenerate a whole periodic block; the latter is usually considerably longer than the former and determines the length of the period. In the oak forests of western France, thanks to favourable seeding conditions, the interval of time between seeding and final fellings over small areas may be only a few years, but the period is fixed at 25 years or more, to ensure complete regeneration of a whole periodic block. The chief factors that determine the length of the regeneration period are:

1. Seed supply: where good seed years are frequent a shorter period is possible than when they occur at longer intervals.
2. Light requirements: for light-demanders the period is short. For shade-bearers a longer period may be used.
3. Hardiness of species: hardy species may be uncovered rapidly and require a shorter period than sensitive species which may require protection by the overwood for many years.
4. Climate: in a mild climate a young crop of sensitive shade-bearers may be uncovered more rapidly than on sites subject to severe frost or drought, so the period may be shortened in mild climates.
5. Soil and vegetation: soil conditions favourable to regeneration (for example, sand and gravel for Scots pine, sandy loam for oak, absence of raw humus or strong weeds for most species) allow short periods. Where the soil is likely to dry out or where strong weeds are feared, the canopy is opened carefully and the period is longer.
6. Control of mixtures: a long regeneration period favours shade-enduring species and a short period favours light demanders.
7. Liability to injury: in Norway spruce damage done during thinning, felling, and extraction causes stem and root rot in young and older trees (see page 46). A long regeneration period increases the risk of damage to young trees and gives more time for the spread of stem and root rot in young and old trees. This does not occur in European silver fir so a long period can be adopted.
8. Risk of fire: in some conifer forests, even when the species are hardy light-demanders, a few seed bearers are retained for many years after the young crop is established, as an insurance against fire, and the period is thereby prolonged.

These factors are considered in relation to each other so that a fair estimate can be made of the time required to obtain complete regeneration and establish the young crop beyond the risk of danger. In Europe the regeneration period for Scots pine is generally from 4 to 10 years. The period generally applied to oak, beech, and Norway spruce, and often also to European silver fir in mixture with spruce varies from 20 to 30 years. Where Silver fir forms the bulk of the crop a long period—40 years or more is frequently adopted, but the system then becomes irregular shelterwood (Chapter 10) and can no longer be called the uniform system.

In theory the periodic block should be completely regenerated by the end of the period. In practice regeneration often does not establish itself as quickly as was hoped, and some areas may still be in the secondary or final stage at the end of the period. There is generally no delay in starting regeneration in the next periodic block while the arrears of fellings in the previous block are worked off separately. Sometimes if a periodic block is not completely regenerated by the end of the period, the remaining seed bearers are felled and regeneration is completed by planting.

It will be apparent, therefore, that the period is an estimate, which may or may not prove accurate, and which frequently has to be departed from in practice. Hence the idea of definite periods of stated length, although it persists under the favourable conditions prevailing in parts of France, has been abandoned, or has never been accepted in some other parts of Europe, particularly in regions where conditions for regeneration are not favourable or where gales or other disasters upset calculations. In the latter case a rough estimate of the time taken to regenerate a given area is usually made for purposes of calculating yield, but the idea of definite periods for the regeneration of periodic blocks does not exist. Compartments allotted for regeneration may already be partly regenerated with advance growth in gaps, and the regeneration period will then represent the time required to complete the regeneration. In some cases the period—not actually stated— varies considerably from place to place, according to the nature of the crop and even within the same compartment, a longer period being adopted where shade-bearers prevail than where most of the crop consists of more light-demanding species.

Nature of periodic blocks

There are considerable differences in the manner in which periodic blocks, if recognized, are constituted. When the uniform system was developed in France in the middle of the nineteenth century the procedure followed was to divide each complete working unit of forest, called a felling series, into as many large self-contained periodic blocks as there were periods in the rotation. These periodic blocks were allocated once for all to its respective period and '*la méthode des affectations permanentes*' is still in force in several oak forests, the beech forests of Normandy, and some coniferous forests because it possesses several advantages. Operations such as fellings, thinnings in pole stage crops, and cleanings in young crops are done over large self-contained areas. This aids supervision of felling and extraction and the concentration of operations into compact areas brings economy of working.

The formation of fixed self-contained periodic blocks and their allocation to all periods of the rotation has the great disadvantage that it cannot be maintained for long in regions subject to gales, snow, insect attacks, and other disasters. From time to time portions of the area are damaged or destroyed and have to be regenerated; this throws out the preconceived arrangement, and if disasters are frequent the periodic blocks lose their identity and have to be abandoned.

The formation of fixed self-contained periodic blocks also creates difficulties for management, because it involves artificial allotment to definite periods of some crops which may be too old or too young for the period to which they have been allotted. They either come under regeneration before their time or they have to be retained until they are past maturity; and in either case, apart from the economic disadvantages, regeneration can be made difficult. The problem of allocating crops to large periodic blocks according to age can be overcome to a great extent by forming scattered periodic blocks; each compartment or even each stand, is considered separately and allocated to its appropriate period. '*La méthode des affectations revocables*' is followed in many European forests and its use has increased. When introducing the uniform system, it may not always be advisable to allot all crops to definite periods in advance, in which case allotment is made to period I (to be regenerated first) and often to one or more other periods in addition.

The formation of permanent periodic blocks, whether self-contained or scattered, is applicable only in regions where the climate is not severe and applies particularly to broad-leaved forests, which are less subject to the risk of serious damage by gales, snow, or insects than conifer forests. In the conifer forests of the mountains of central Europe where severe gales do occur, fixed periodic blocks are not recognized. Instead when a working plan is revised, the areas to be regenerated during the interval between that revision and the next are decided according to the state of the crop and marked on the map. Regeneration fellings are carried out on these areas and by the time the next revision takes place, say 10 years later, some areas may have been completely regenerated, while others are still under regeneration. The latter will be retained as regeneration areas under the revised working plan, while new areas are taken up for regeneration and added to them in place of areas already regenerated. The same procedure is followed at each revision so that there are always areas under regeneration, although these are periodically changing. This amounts to what is called a floating periodic block. '*La méthode du quartier du régénération*' was introduced into some of the coniferous forests of the French Jura and Vosges by Melard in 1894 to replace fixed periodic blocks. Here it is called the '*Quartier bleu*' method because areas under regeneration for the time being are marked blue on the map; areas to be regenerated next are coloured yellow (*Quartier jaune*) while the rest of the forest remains uncoloured and is called the '*Quartier blanc*'.

Even where no serious disasters are feared, fixed periodic blocks may be impracticable because natural regeneration may not always be obtained when and where it is wanted, although in some places it may appear spontaneously in great abundance without deliberate aid from the forester. In such cases the practical procedure is to 'follow the regeneration' rather than attempt to force its appearance. In the tropics where much has still to be learned about the silviculture of many species, the principle of the floating periodic block is especially appropriate and has been applied to some of the *Shorea robusta* forests of India (see page 158).

Where regeneration takes place rapidly and regularly, periodic blocks are sometimes dispensed with and an arrangement of annual coupes may be followed similar to that of the clear cutting system. Thus in some Scots pine forests of Normandy in France, seeding fellings are made in annual coupes by area, regeneration generally establishes itself readily, and a few years after the seeding felling the seed bearers are removed in one or two secondary fellings (see page 118). The yield is fixed entirely by area, but the annual volume out-turn is reasonably constant because the secondary fellings follow the seeding with sufficient regularity.

To sum up, the methods of adapting the uniform system to periodic blocks or otherwise are:

(1) fixed periodic blocks, which may be self-contained or scattered;
(2) floating periodic blocks, usually scattered;
(3) annual coupes by area.

Numbering periods and periodic blocks

When periods and periodic blocks are fixed, the practice in France is to number them I, II, III etc. and to retain these numbers permanently. Thus in the oak forest of Bellême (Orne) when this system was introduced in 1856, the rotation was fixed at 200 years and divided into eight periods of

25 years each. This has been adhered to ever since, so that in 1977, for instance, the ages of the crops in the periodic blocks already regenerated were theoretically as follows:

Period		Age of regenerated crop in 1977 (years)
I	1856–1880	97–121
II	1881–1905	72–96
III	1906–1930	47–71
IV	1931–1955	22–46
V	1956–1980	up to 21

The year 1977 was the 22nd year of the fifth period and in that year the crops in periodic block V should have been in the final stage or already regenerated. In the ordinary course the remaining periodic blocks would be due for regeneration during subsequent 25-year periods, from period VI (1981–2005) to period VIII (2031–2055).

Effect of regeneration period on form of crop produced

The aim of the uniform system is to produce young crops that are as even-aged as possible. The length of the regeneration period has a decided effect on the form of the young crop. Short periods produce even-aged crops, but the longer the period the more uneven-aged or irregular the crops become. A stage is reached when the uniform system gives place to the irregular shelterwood system (Chapter 10). Ordinarily the dividing line is reached when the period is between 30 and 40 years, although the form of the young crop rather than the length of period is the best criterion. With periods up to 30 years, unevenness in the young crop due to differences in age tend to disappear during the pole stage, and by the time the timber stage is reached the crops have the appearance and character of even-aged crops, with long, clean, cylindrical stems.

Uniform system with artificial regeneration

The uniform system with artificial regeneration is sometimes used to introduce new species or provenances, or to increase the proportion of a particular species already existing but not in sufficient quantity. The young crop is introduced by direct seeding or planting. The procedure is otherwise similar to that followed with natural regeneration, the canopy being opened gradually or rapidly according to the light requirements and hardiness of the young crop and the risk of frost, drought, strong weed growth and other factors. Generally a short period may be adopted since a sown or planted crop can be established rapidly. Artificial regeneration under a shelterwood system is used most commonly to introduce sensitive shade-bearers or to obtain a mixture by combining artificial regeneration of an introduced species with natural regeneration of existing species. Under some conditions species that are moderately light-demanding may be introduced under a shelterwood, as in the case of oak in the Spessart highlands of West Germany.

Advantages and disadvantages of the uniform system

The advantages and disadvantages of shelterwood systems in general (page 91) apply also to the uniform system. The advantages of the uniform system when compared to other shelterwood systems are:

1. The fellings are simpler to carry out than in most shelterwood systems. The seeding felling is a straightforward, even opening of the canopy over large areas, in the manner of a thinning.
2. Crops of even-aged type are produced, typically with long, clean stems.

The disadvantages are:

1. Damage to young growth during harvesting can be greater than in the strip or wedge systems, although it can also be greatly reduced if care is taken and the work is well organized.
2. Seed bearers isolated in seeding or secondary fellings are liable to be thrown by wind, especially if attention has not been given to stem taper or crown and root development during thinnings. The spruces and other shallow-rooted species are particularly prone to windthrow.
3. On hot sites, isolated seed bearers of thin-barked species, such as beech, are liable to suffer from sun scorch (Peace 1962).
4. In localities subject to frost or drought, the overhead cover provided by the uniform system is less effective than the side protection provided by the strip systems.
5. Where damage by wind and snow is prevalent the regular crops produced by the uniform system are more subject to damage than the more irregular crops produced by the irregular shelterwood, and selection systems.

Generally speaking, where the climate is not severe, and conditions of seeding and regeneration are favourable, the simplicity of the uniform system gives it special advantages over most other systems. Where serious damage by gales or snow is to be feared, or the supply of seed is uncertain, or the soil conditions impede root development, the uniform system is not much used.

Treatment of individual species and mixtures

Considerable variation in treatment is needed to provide for differences in the silvicultural characters of different species. Regenerating mixtures is further complicated because it is exceptional to find two species regenerating with equal freedom and vigour; in most cases special measures must be taken to hold the balance between the two and, if necessary, assist one against the other. Where one species demands more light than the other, the course most commonly adopted is to open the canopy slightly and obtain partial regeneration of the more shade-enduring species, then open it further by removal of seed bearers of that species so as to allow the more light-demanding species, to regenerate or fill the gaps. If there is a third, and still more light-demanding species in the mixture, a further opening of the canopy is made subsequently, seed bearers of the light-demander being retained somewhat in the manner of standards, to regenerate any blank spaces remaining.

However, this procedure has sometimes to be modified. Where one species tends to become dominant at the expense of the other, it is generally necessary to establish the latter first in groups around seed bearers of the species concerned, or even to introduce them artificially to the extent desired. The dominant species is regenerated afterwards by uniform opening of the remainder of the canopy.

Beech is a shade-bearer with a heavy seed which does not spread naturally to any distance from the parent tree. Beech masts occur more irregularly than most other European species; warm, sunny periods in June and July which favour the initiation of flower buds in one year must be followed in the next year by a flowering season free from frost plus summer weather favourable to development and ripening of the mast (Matthews 1964). Full mast years occur on average every 5 or 6 years in mild climates, such as those of Denmark, western France, and southern England and every 8 to 12 years in most parts of Germany, southern Sweden, and northern Scotland, apart from occasional partial mast years. At higher elevations beech does not seed more than once in 15 years. In Europe partial seeding can be used to a greater extent than in the case of more light-demanding species, and patches of partial regeneration can be a nucleus for more complete stocking. In southern England however, Watt (1923) found that the hazards from drought and predation by animals were so formidable that partial masts rarely have any practical result.

The height increment of beech seedlings is slow at first; they are sensitive to frost and drought, and do not tolerate acidity or wetness caused by impeded drainage in the soil (Brown 1953). A thick layer of undecomposed leaves is unfavourable in preventing the radicle from reaching mineral soil. As already noted (page 101) conditions for regeneration are favourable when a light growth of herbaceous or shrubby vegetation is beginning to appear; in Britain one of the best indicators of favourable light and soil conditions is Wood sorrel, *Oxalis acetosella*. A thick matted growth of grass, such as may follow too heavy an opening of the canopy prevents regeneration. Wavy hair grass, *Aira flexuosa*, which is sometimes common in beech forests, is one of the worst grasses in this respect. In southern England, bramble, *Rubus fruticosus*, competes strongly with beech seedlings and must be controlled throughout the period of regeneration (Brown 1953).

As a rule the regeneration period varies from 20 to 30 years. Under favourable conditions seeding fellings are made irrespective of the occurrence of seed years. Where seed years must be awaited, and the flowering and early stages of fruit formation in spring and early summer indicate a good mast, a cutting plan is prepared during summer so that as large an area as possible may be marked for felling during the following winter.

Generally more than one seeding felling is needed to effect complete regeneration. The seeding felling is light, the canopy being opened only to the extent that the crowns touch when swayed by wind. In the beech forests of France 25 to 35 per cent of the crop is removed, leaving about 170 stems/ha under average conditions (Fig. 11). Any understorey of shrubs and suppressed trees should be removed. The secondary fellings begin when the young crop is knee-high and 6 to 8 years old; they should be continuous and gradual. Under the most favourable conditions there are generally at least three, including the final felling. Trees with large spreading crowns should be removed first as they deprive the seedlings of water and nutrients thus preventing regeneration. The final felling should not be delayed longer than is necessary. As already stated the bark of isolated beech trees tends to become scorched by the sun (Peace 1962); damage done during harvesting is intensified when the young crop is allowed to grow too tall.

Hornbeam is a frequent companion of beech in French forests. It seeds freely every 2 or 3 years and springs up in great profusion; to favour the beech it is necessary to keep the canopy dark enough during the seeding stage to kill off hornbeam seedlings which stand less shade than those of beech.

The **oaks** are light-demanders; the acorns are heavy and do not naturally spread far from the parent tree. The periodicity of good mast years varies greatly in different parts of Europe. In the mild climate of the Adour valley in south-west France, annual seeding of Pedunculate oak is the rule. In England the periodicity is from 3 to 5 years for both Pedunculate and Sessile oak, while Normandy and western and central France generally have good mast years of Sessile oak every 4 to 8 years. In the more severe climate of north-eastern France they are much less frequent; on the plateau between the rivers Moselle and Meurthe in Lorraine, oak seeds well on average only at intervals of about 20 years, and under such conditions the uniform system is ruled out. In the Spessart highlands of West Germany full mast years of Sessile oak occur on average every 10 or 12 years.

Under the favourable conditions of western France, complete regeneration has been obtained in 6 to 10 years, this being the actual interval between the seeding and final felling. Nevertheless, the regeneration period for whole periodic blocks is sometimes much longer because it takes some years for seeding fellings to be completed over a whole periodic block.

As a rule about one-third of the crop is removed in the seeding felling, leaving about 75 to 120 seed bearers/ha, according to age and size, and there is a space of several metres between crowns. The stems are on average 10 m apart. All undergrowth and low cover should be removed. In France seeding fellings are generally made irrespective of mast years. When a good mast year occurs the ground is thickly carpeted with seedlings, in which case the secondary fellings follow rapidly (Fig. 13). Generally speaking there are two secondary fellings, including the final felling, but under favourable conditions a single final felling is done after the seeding felling. Oak seedlings are less sensitive than those of beech, but they do suffer in frosty localities, so where frost is feared the canopy is opened more cautiously, both in seeding and secondary fellings.

The failure of natural regeneration in most British oakwoods was explained by Watt (1919). The main factors causing failure are predation of acorns by animals and birds, and damage to the foliage of seedlings by Oak mildew (*Microsphaera alphitoides*) and larvae of the Tortricidae. Successful regeneration is most likely in moist oakwoods and least likely in dry oakwoods. Lanier (1986) also stresses the high loss of seedlings from these and other causes which endangers natural regeneration of oak.

Mixtures of oak and beech are common in Europe; it is exceptional to find an oak forest without some beech which is kept partly to protect the soil. In localities especially favourable to oak this species usually holds its own against beech, and a successful mixture can generally be obtained by making regeneration fellings in the way described for oak. Where, as

Figure 13.   Shelterwood uniform system with natural regeneration of oak after a seeding felling. The oak seed bearers have clear stems 15 m long. Note prolific regeneration. Forest of Réno-Valdieu, Normandy, France.

frequently happens, beech regenerates with greater freedom and tends to outgrow and suppress the oak in youth, care should be taken during thinnings to develop a large percentage of good oak seed bearers. It is also advisable to wait for an oak mast year before carrying out the seeding felling, which should be made open enough to effect oak regeneration. Removal of the overwood in secondary fellings should proceed as rapidly as possible, and if necessary young beech should be cut back and topped where it tends to suppress the oak. Where oak mast years are very infrequent the uniform system with natural regeneration is ruled out.

Sessile oak and beech both occur naturally in the Spessart highlands of Bavaria, but the proportion of oak obtained by natural regeneration is insufficient and special measures are taken to increase the quantity of this species at the expense of beech, which regenerates freely, grows more rapidly in youth, and would oust oak if permitted to do so. The crop is opened out widely, about two-thirds of the growing stock being removed, and Sessile oak is introduced by sowing acorns under the light cover of the open beech crop, which is retained to protect the young seedlings against frost and drought, retard weed growth, and promote a mixture of beech by natural regeneration. The beech overwood is removed at intervals of 2 to 3 years, the final felling being done within 8 years of the seeding felling and sowing of the oak. From the time oak sowings are done the young beech has to be uprooted or cut back where necessary to prevent it from becoming dominant. The regeneration of a whole compartment is completed by this method in 20 to 30 years, involving as a rule 2 or 3 beech seed years.

Scots pine is a strong light-demander with a winged seed which is carried in quantity by wind to a distance from the parent tree equal to twice its height (Steven and Carlisle 1959; Booth 1984). Some seed is produced almost every year and good seed years occur at intervals of 3 or 4 years. Good crown development is important in the production of an abundant supply of cones. Seedlings of Scots pine are hardy to frost and drought but will not tolerate shade. Regeneration usually appears on sandy or gravelly soils; in dry places a moderate cover of heather *Calluna vulgaris*, or low open bracken *Pteris aquilinum*, or broom *Sarothamnus scoparius*, is a help rather than otherwise, in preventing excessive drying of the soil. Regeneration is greatly stimulated by removal of a layer of needles and exposure of the mineral soil (Edwards 1981; Jones 1947). Scarifiers are often used in many parts of Europe to create favourable seed-beds for Scots pine regeneration (Low 1988).

Scots pine is well adapted to the uniform system provided care is taken to develop the crowns and roots of seed bearers by regular thinnings at intervals of not more than 5 years; failing this, the trees when isolated by the seeding felling are apt to be thrown by wind. In applying the uniform system the procedure is simple. The seeding felling consists of a wide spacing of seed bearers, and a final felling should follow when the young crop is not more than 30 cm high, that is, within a few years of the seeding felling (Fig. 14).

Norway spruce is a moderate shade-bearer with a light, winged seed which is carried by wind to some distance from the parent tree. Seed years are somewhat irregular; they occur on average at intervals of 3 to 10 years, but even longer intervals can occur. The seedlings are fairly frost-hardy but are sensitive to drought owing to their shallow root system. In moist soils they sometimes spring up in great abundance. They do not tolerate heavy shade, nor do they recover satisfactorily after suppression. In central Europe raw humus is one of the greatest obstacles to regeneration; removal of the undecomposed layer of needles produces conditions favourable to germination and the establishment of seedlings.

Figure 14. Shelterwood uniform system with Scots pine. Seeding felling showing good distribution of seed bearers and a clean forest floor. Forest of Bord,
France.

The uniform system is unsuitable for Norway spruce in localities subject
to gales, since this shallow rooting species, when isolated in regeneration
fellings, is too liable to windthrow. For this reason, if the uniform system is
to be applied with success it is essential that the conditions should be
entirely favourable (see page 95), so that the young crop may be established rapidly. The seeding felling should be done in a good seed year and it
should be made cautiously to avoid isolating the trees and exposing them to
wind damage. At the same time sufficient light should be provided to
promote regeneration; this can be done by removing one-quarter to one-
third of the growing stock. The final felling should be done within 4 or 5
years, with or without an intervening secondary felling, and blanks should
be filled by planting.

In central Europe the uniform system for Norway spruce has largely been
replaced by systems in which uniform fellings sometimes play a part. Here
Norway spruce is commonly treated in mixture with European silver fir,
with or without beech, Scots pine, and European larch.

European silver fir is a strong shade-bearer with a heavy winged seed,
which is conveyed by wind to distances equal to the height of the parent
tree. A good seed year occurs on average every 2 to 10 years according to

locality. Under favourable conditions seed production is frequent and regular. The cones are borne high in the crowns of mature trees and best regeneration is obtained from trees 90 to 120 years of age. The seedlings are sensitive to frost and drought and are readily browsed by deer; they stand a considerable amount of shade and have strong power of recovery after suppression.

Silver fir can be treated successfully under the uniform system. There are often quantities of seedlings on the forest floor before any seeding felling is done and the secondary fellings can begin forthwith. Where a seeding felling is done it should consist mainly of removing any understorey and raising or heightening the canopy, but with only a slight opening of it. The secondary fellings should be done cautiously, and there should be several of them extending over several years. The young crop requires protection till it is well-established and the cover should be removed gradually.

Norway spruce and European silver fir with other species in mixture require a very cautious seeding felling to allow the Silver fir to establish itself. If beech is present it tends to appear in groups around beech seed bearers. A few years later the canopy is opened sufficiently to enable Norway spruce to regenerate, spruce and Scots pine seed bearers are retained, and Silver fir and beech are removed where possible. As soon as Norway spruce has regenerated itself sufficiently, the canopy is opened further in one or more secondary fellings, scattered Scots pine seed bearers being retained during the final stage to seed up any gaps. The mixture can be regulated by the rate at which the canopy is opened. A gradual opening encourages the Silver fir and beech, and a more rapid opening encourages the Norway spruce and Scots pine.

In some forests of the Vosges and French Jura, Norway spruce and Silver fir in mixture are treated successfully under the uniform system. Regeneration of Silver fir often appears in abundance before the canopy is opened and persists for some time under the heavy shade. A seeding felling is then made to stimulate the spruce, followed by two or three secondary fellings, including the final felling. Regeneration is thus completed within 20 years, although the period laid down for regenerating whole periodic blocks may be as much as 30 years or more.

While it is desirable to complete the regeneration fellings as rapidly as possible owing to danger from wind, care is necessary to avoid too sudden an opening of the canopy, which may result in the death of pre-existing Silver fir seedlings. Groups of advance growth of Silver fir are retained, so the young crop often appears irregular for a time. Where deer are prevalent Silver fir regeneration may fail to make headway until Norway spruce has established itself and formed a protective cover for the fir.

Application in practice

Systems of regeneration resembling the uniform system as now practised appear to have been known over 400 years ago. In Germany records dating from the end of the fifteenth century refer to regeneration over large coupes, producing regular crops. In the forest ordinances of the fifteenth and sixteenth centuries, rules stated the number of seed bearers to be retained so as to secure natural regeneration. Many writers during the eighteenth century refer to natural regeneration obtained by leaving seed bearers. In Hungary an ordinance of 1565 prescribed, among other matters, the order of fellings and reservation of seed bearers.

In France the idea of coupes by area, leaving a specified number of trees for seeding, took shape at least as early as the fourteenth century and led subsequently to the adoption of a system called *tire et aire*. The latter had points of resemblance to the uniform system which began to be introduced in place of it before the middle of the nineteenth century. It is clear that in France something closely resembling the uniform system of today was thought of, if not actually practised, at least 400 years ago.

The uniform system in its modern application will always be associated with the name of Georges-Ludwig Hartig (born 1764, died 1837), who as Director-General of the Prussian State Forest Service from 1811 to 1837 greatly influenced the silviculture of his time and strove to introduce regular order everywhere. The first edition of his 'Anweisung zür holzzucht für förster' appeared in 1791 and marked a new epoch. Hartig, acting on experience gained in oak and beech forests, advocated even-aged regeneration over areas as little scattered as possible, so as to improve the supervision of fellings. He recognized three regeneration fellings—a seeding felling with the crowns of the seed bearers almost touching each other; secondary felling, the canopy being opened to 15–20 paces between crowns, when regeneration is more or less complete and 25–40 cm high; and the final felling removing the remaining seed bearers when the young crop is 60–120 cm high. Heinrich Cotta (born 1763, died 1844) in the first edition of his 'Anweisung zum Waldbau' (Cotta 1817) recommended a procedure similar in the main to that of Hartig.

In 1827 Bernard Lorentz (born 1775, died 1865), who was Director of the Forest School at Nancy, recommended the introduction of Hartig's uniform system into France. Following success at Bellême (Orne) it was used in the neighbouring forest of Réno-Valdieu in 1827, but was not applied on a large scale until the middle of the century, when it was introduced in oak and beech forests in place of the old *tire et aire* and in some

Norway spruce and Silver fir forests of the Jura. Parade (born 1802, died 1864), who had been a pupil of Cotta at Tharandt, was another strong supporter of the uniform system. Now the uniform shelterwood system is used to regenerate 90 per cent of oak and beech forests in France (Oswald 1982).

The uniform system is less used for beech and oak in Germany today. It is applied to beech in Denmark, and in Scandinavia is widely used to regenerate Scots pine and Norway spruce, especially in Sweden (Hagner 1962) and Finland. During the later years of the nineteenth century the uniform system was successfully applied to Scots pine in the valley of the river Spey in north-east Scotland. Some use has been made of it elsewhere in Britain but Scots and Corsican pine are almost wholly regenerated by the clear cutting system.

In the United States of America the uniform system has proved useful for Eastern white pine, in the cherry-maple forests of the Allegheny mountains and the mixed broad-leaved forests of the Appalachian mountains. All these are eastern forest types. Elsewhere it is used in regenerating Ponderosa pine and several species of southern pines either as the shelterwood uniform system or the seed tree method (Burns 1983).

The seed tree method

General description

This is a form of the uniform system in which the seed bearers are widely spaced over the area to be regenerated. It is mainly used for light-demanding species with seed that is dispersed by wind, such as *Pinus roxburghii* in northern India and Pakistan, Loblolly, Longleaf, Slash, and Shortleaf pines in the southern United States, and Ponderosa pine and Western larch in the Western states of America.

All the mature timber is removed from a coupe in one felling save for a small number of seed bearers, rarely < 10 or > 25/ha, which are left standing singly, in lines or in small groups to regenerate the new crop. Sometimes the seed bearers are retained to put on increment as well as produce seed.

The success of the seed tree method depends on:

(1)　careful choice of seed bearers for phenotypic quality of stem and branching habit, absence of serious damage by disease, evidence of ability to produce seed, and windfirmness;

(2) high production of viable seed per tree;
(3) adequate dispersal of seed on to well-prepared seed-beds;
(4) good survival of seedlings during the critical early stages of growth.

Windfirmness is a primary consideration in choosing seed trees. The sudden removal of the protection provided by the whole stand and their isolated condition renders them particularly susceptible to windthrow. The most windfirm trees are usually the healthy dominants with stems that taper strongly, deep live crowns, and correspondingly strong root systems (Smith 1986). It is very desirable to release the crowns of the seed bearers in preparatory fellings before the main crop is harvested. Fertilizer may be applied to stimulate development of foliage, flower buds, and seed. A common prescription for pine species is 100 kg of nitrogen, and 50 kg each of phosphorus and potassium per hectare. A final felling to harvest the seed bearers is made whenever possible, otherwise they are left on the site.

The seed tree method is not suitable for shallow rooting species like the spruces nor is it applicable to any species growing on wet or thin soils which restrict rooting to the upper layers.

Application in practice

In the north-west Himalayas this form of the uniform system has been applied to *Pinus roxburghii* for almost a century. This gregarious tree, which is strongly light-demanding, forms extensive forests on hot dry slopes, chiefly between 600 and 1700 m elevation. The average periodicity of good seed crops varies from 2 to 3 or 4 years according to locality. The winged seed is borne by wind to some distance from the parent tree and, given a suitable distribution of seed bearers with good crowns, plus controlled burning to remove a thick layer of undecomposed needles and competing vegetation, natural regeneration usually springs up in abundance, especially on well-drained porous soils where the roots of the seedlings can develop well. On stiff clays and shallow soils overlying limestone, the roots cannot grow sufficiently and regeneration may succumb to drought on hot aspects (Troup 1952, p. 56).

Many of these forests are burned regularly and used for grazing, so strict protection is required to keep regeneration; ground cultivation may be needed where the soil surface has been compacted by the animals (Champion *et al.* 1973). In general, 12 to 15 seed bearers/ha are retained on the cooler aspects but on the hotter south and south-west aspects 20 to 25 are left. Seedlings appear after a good seed year and the area is strictly protected from burning and grazing until this regeneration is 1–2 m tall on gently sloping ground and 2.5–3 m tall on steep slopes. The seed bearers are retained until the regeneration is 6–7 m tall because the risk of accidental fire is always present.

The typical variety of Slash pine (*Pinus elliottii* var. *elliottii*) occurs on several million hectares of commercial forest land in the southern United States (Shoulders and Parham 1983). It is widely planted but is also regenerated by direct seeding and by the uniform system, including the seed tree method. Slash pine grows best on deep, well-aerated soils that supply plenty of moisture during the growing season. The climate comprises relatively mild winters and long, warm, humid summers. Most rain falls during February to March and July to August. Slash pine is considered to be generally light-demanding. If fire were absent it would be replaced by mixed broad-leaves.

Good seed crops appear about every 3 years, with some seed in most years. Cone production may be greatly increased by stem wounding, by fertilizer application before flower buds are formed, and by crown development following heavy thinning. Seed is dispersed during October and 90 per cent falls within 25 m of the parent tree. In the forest, germination takes place from November to April. Seed germination and survival of seedlings may be improved by exposing mineral soil prior to seed fall; the broad-leaved understorey must be controlled at the same time.

Usually, controlled burning is done during the winter before a good crop of seed is expected, to reduce the understorey and prepare the seed-bed. The main crop is felled leaving 15 to 25 seed bearers/ha. The critical stage for the regeneration is during the first summer following germination when seedlings can be lost through drought. The seed bearers must be harvested within 3 to 5 years to promote vigorous growth of the new crop and reduce the amount of damage during felling.

The seed tree method has also been applied successfully to Ponderosa pine on the western slopes of the southern Cascade mountains and Sierra Nevada where this light-demanding species forms almost pure, even-aged stands (Oliver *et al.* 1983). The rainfall is 760 to 1720 m and falls mainly from November to April. The growing season is warm, dry, and mostly cloudless.

Moderate to good seed crops appear at intervals of from 1 to 3 years. Ninety per cent of seed falls within distances equal to 1.5 times the height of seed bearers. The seed-bed must be treated to expose mineral soil. The main crop is felled leaving 10 to 20 and sometimes 30 seed bearers/ha. Large seed bearers compete strongly with the regeneration for soil moisture and nutrients which retards the growth of young plants at distances of 12 m or even more. The mature trees are therefore removed as soon as satisfactory stocking has been achieved. On average 8 per cent of the regeneration is lost through damage during felling of seed bearers and skidding the logs with tractors. Direct seeding is an expensive alternative to natural regeneration and planting is most commonly used to restock stands of Ponderosa pine.

9

The group system

Fr. Régénération par groupes; Ger. Gruppenschirmschlag; Span. Cortas por bosquetes uniformemente repartidos

General description

The group system like the uniform system, produces crops that are of even-aged type by the time the pole stage is reached. During the thinning stages the treatment of crops under both systems is identical, and it is not until the regeneration stage is reached that they differ. Under the group system the first step is to go over the compartment and search for any promising groups of advance growth which may have appeared in gaps caused by wind, snow, or other agencies. If these groups of advance growth require freeing, the gaps are widened by felling trees around their edges (Ger. *Umsaumungshieb*) while at the same time, or as soon as a seed year occurs, a seeding felling is made in the form of a ring round each gap (Ger. *Rändelhieb*). If there are not sufficient natural gaps over the compartment further gaps are created by felling trees in small groups. In this way numerous gaps are distributed over the regeneration area, which may be a whole compartment, or more commonly, only a portion of a compartment at a time.

In the Norway spruce and Silver fir forests of Bavaria, when the initial gaps were made the general rule was to fell the largest trees, leaving moderate-sized trees to act as seed bearers and put on increment; these moderate-sized trees were found to be better seed bearers than the larger ones.

As soon as regeneration has appeared in the artificially created gaps, a seeding felling is made around each gap as before. Regeneration thus spreads centrifugally round each gap; secondary and final fellings in turn follow the seeding fellings, while new seeding fellings continue outwards into the unopened old crop in ever-widening circles. The groups of regeneration thus become larger and larger and eventually meet, when the last remaining seed bearers separating the various groups are removed and

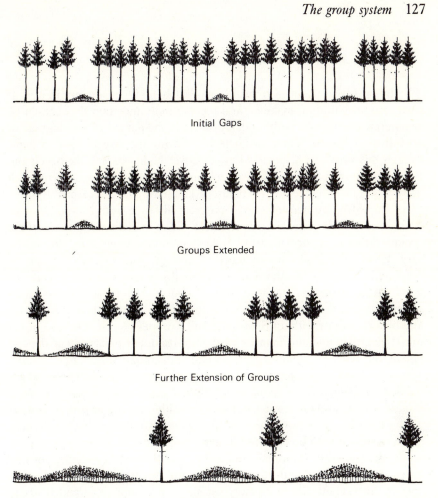

Initial Gaps

Groups Extended

Further Extension of Groups

Final Stage

Figure 15. Group system showing successive stages of regeneration.

only the regenerated young crop remains. The process of regeneration is shown in Fig. 15. Regeneration is sometimes aided by scarification in and around the clearings. After the final felling blanks are planted up.

In freeing groups of advance growth at the beginning of the regeneration fellings, it is necessary to concentrate on advance growth that is not seriously suppressed, and to remove any that has had its vigour destroyed by suppression. Some shade bearers, notably Silver fir, have the capacity to recover after moderate suppression, but Norway spruce and more light-demanding species like Scots pine and European larch do not recover satisfactorily if their vitality has been seriously impaired.

Forming gaps and regulating mixtures

The initial gaps produced by wind or other natural causes vary in size. If their size or condition make natural regeneration uncertain they should be regenerated artificially without delay owing to the danger of soil degrade or an influx of weeds. There is no fixed rule for the size of cleared gaps, though this varies with species. For sensitive shade-bearers like beech and Silver fir the canopy is only just broken, and subsequent enlargement of gaps is gradual; hence to regenerate a compartment in reasonable time, there should be numerous gaps at no great distance from each other. For Norway spruce the gaps are commonly 18–23 m in diameter or more. If the group system is applied to strong light-demanders, larger gaps are necessary. Where there are two or more species present, the mixture can be regulated effectively by attention to the initial size and subsequent enlargement of the clearings. In those parts of central Europe where the group system is used, the forests consist of a mixture of Norway spruce, Silver fir, and beech, with or without Scots pine and European larch. The size of the initial clearings and the manner of their subsequent enlargement depend chiefly on whether shade-bearers or light-demanders are to be favoured. Where Norway spruce tends to become dominant and special measures are needed to favour Silver fir or beech, initial gaps of small size are made around seed bearers of these latter species, with only one or two, or at most a few, trees being removed so that the canopy is only just broken. Where Norway spruce is to be favoured larger gaps are made. Where one species tends to regenerate vigorously and to become dominant, seedlings of that species are cut out in the gaps where they threaten seedlings of other species. In regions where Norway spruce is at its optimum and tends to become dominant—as in the Böhmerwald in Czechoslovakia—this practice is sometimes needed to maintain an admixture of Silver fir and beech. Where there are strong light-demanders in the crop, such as Scots pine or European larch, scattered seed bearers of these are left standing for a time over the young crop after regeneration of other species is completed, so as to seed up blanks.

Artificial regeneration may play an important part in regulating mixtures. When the proportion of beech is not sufficient, it may be introduced by planting in groups in advance of the regeneration fellings, after making small gaps in the canopy. This procedure is used in the pure crops of Norway spruce planted in central Europe during the nineteenth century. On patches of good soil, groups of oak, ash, or maple are sometimes introduced in the same way. In this case clearings of some considerable

size are made, particularly for oak which is introduced in groups sufficiently large to form self-contained units to be left over for a second
rotation.

Protective measures

To guard against damage by wind, a frequent modification of the procedure
just described is to begin regeneration fellings in the eastern portion of the
compartment and continue them in a westerly direction, so that the unopened stand is retained for as long as possible on the westerly or dangerous
side of the group felling. For this purpose it is usual to divide the compartment into, say, three parts—eastern, western, and central. Regeneration
fellings begin first in the eastern, later in the central, and last in the western
part. It is usual to make the clearings close to each other so as to hasten
completion of regeneration in each portion, otherwise the time taken to
regenerate the compartment will be unduly long.

In places where the soil is liable to dry out when exposed to the sun, a
procedure adopted in parts of Czechoslovakia is to enlarge the gaps successively on the south side only; this provides side shade from the sun on the
southern edge of the gap where the seedlings are youngest and therefore in
their most critical stage. This procedure is more effective than making
group fellings progressively from north to south, as is sometimes done. On
hillsides it is usual to begin regeneration at the top of a slope and continue
by degrees in a downhill direction so as to reduce the damage done during
harvesting. A combination of the strip and group systems has been evolved
to give effect to these protective measures (see page 145).

Periods and periodic blocks

Under the group system the maximum duration of the period is similar to
that of the uniform system, but the minimum duration is longer; periods of
20 to 30 years are the rule in forests where Norway spruce is the principal
species although periods up to 40 years are sometimes adopted. Frequently
the period is not definitely fixed but only approximately indicated. A long
regeneration period is usually objected to for Norway spruce because
damage done during felling and extraction produces stem and root rot in
the young trees and the damage accumulates with time (see page 46). In
Europe the group system has generally been used in regions subject to gales
and snow, hence floating periodic blocks are the rule; fixed periodic blocks
are not suitable for these conditions.

Form of crop produced

The young crop produced by the group system results from a succession of seed years and so is somewhat uneven-aged, with a wavy profile due to the presence of numerous thickets of different ages (see Fig. 15). If regeneration proceeds slowly and the clearings are widened gradually, the uneven-aged condition is more pronounced than where regeneration proceeds quickly and the gaps are widened rapidly; in the latter case the form of the young crop differs little from that obtained under the uniform system. Where the periods are equal, the group system produces more uneven-aged crops than the uniform system, owing to the inclusion of advance growth, but even under the group system any irregularity generally disappears by the time the pole stage is reached, and this system may therefore be regarded as essentially an even-aged system.

Advantages and disadvantages

The chief advantages and disadvantages of shelterwood systems in general (see page 91) also apply to the group system. The advantages that apply more particularly are:

1. By making use of advance growth a start of several years is obtained in the process of regeneration. This is a distinct advantage where seed years are not frequent, or where regeneration is not always certain.
2. The young crop develops in a more natural way than in the uniform system, and its uneven-aged condition up to the pole stage is considered to give protection against snowbreak and sliding snow on steep hillsides.
3. So long as the gap is small the young crop gains some of the advantages given by side protection combined with overhead light at the early stage (but see under the disadvantages below).
4. Damage by felling is avoided during the early stages of regeneration by directing the felling away from the groups of young growth.
5. Mixtures can be easily regulated or created by regenerating or introducing species in groups, the size of the clearings being adjusted to suit the light requirements and hardiness of the species concerned.

The disadvantages are:

1. Although the soil in the gaps is protected from the sun by the side shade of the old crop during the early stages of regeneration, as the

gaps are enlarged the sun beats down on their northerly edges (in the northern hemisphere). Where the soil is of a dry type this may cause heavy mortality among young seedlings, particularly of shallow-rooted species like spruce, and may thus seriously affect regeneration on those parts of the gap exposed to the sun. In the northern hemisphere this can be countered to some extent by enlarging the gaps in a southerly direction only.

2. Serious damage by wind is liable to occur, especially on the margins of small gaps.
3. Control is difficult, because of numerous, scattered and small centres of regeneration each having to be located at frequent intervals during the course of regeneration. Harvesting also becomes difficult in the later stages as the groups coalesce.
4. Difficulty is sometimes experienced in opening out the gaps sufficiently to keep pace with yield requirements and, at the same time, ensure regeneration.

Application in practice

By the early 1960s the area being clear felled and replanted in British Forestry Commission forests was beginning to rise as the older plantations of spruce reached the age of maximum mean annual increment. The occurrence of windblow in stands approaching 16–18 m top height stimulated more research into its causes and ways of countering it; in addition the appearance of natural regeneration in clearings created by windthrow aroused speculation about its possible use when restocking was done. Neustein (1965) laid out a trial in the forest of Ae near Dumfries in south Scotland to determine the stability of the margins of clearings made in stands of Sitka spruce aged between 32 and 35 years and with top heights of 16–17 m growing on peaty gleyed podsols. The size of clearings were 0.04, 0.10, 0.40, and 4.05 ha. The trees were felled during the winter of 1961–62 and the clearings were sited on comparable soils in places where the stand margins would have similar characteristics. There was a consistent reduction in run of wind with diminishing size of clearings so that the edge trees of the smaller gaps were apparently subjected to less wind; but this was not reflected in a smaller number of windthrown trees when expressed on an area basis because the greater length of perimeter per hectare outweighed the apparent reduction in wind speed.

In 1966, McNeill and Thompson (1982) began to monitor the occurrence of Sitka spruce seedlings in the smallest size of gap (0.04 ha) and their survival in relation to their position in the clearings and the vegetation competing with them. Peat covered most of the ground to a maximum depth of 7.5 cm but mineral soil was exposed in places where spoil from open drains occurred on the surface. The average annual rainfall for the area is 1375 mm and it lies between 240 and 270 m elevation. Within a series of quadrats 1 m square, spaced at intervals of 3 m across the clearings and for 15 m into the old crop, all seedlings were labelled and counted at monthly intervals during the growing season and less frequently in winter, until 1972.

Most of the seedlings studied appeared in 1964, 1968, and 1970. Losses during the first and second seasons were high. On the edges of a clearing seedlings were more plentiful than in the centre or well under the old crop. There was a tendency for seedlings to survive best on shaded quadrats on the south side of a clearing. Two principal causes of death of newly germinated seedlings were heat and drought between June and August, and competition from the ground vegetation. During dry spells of up to 14 days the layer of needle litter dried out considerably and seedlings succumbed. Defoliation by *Elatobium abietinum* also caused considerable losses (see page 42), but well-established seedlings were able to continue growing. Survival was also better on mineral soil than on the peat. Between 1966 and 1972 windthrow gradually extended the perimeter of the clearings; light conditions improved and so also did the height growth of seedlings in the centre of the gaps. However, seedlings under the canopy of the old crop soon became moribund and died. Despite losses due to the causes described, there was ample regeneration to form a new crop (the densities ranged from 20 000 to 300 000/ha) and the best plants had reached a height of 1 m after six growing seasons.

Low (1985) reported the occurrence of natural regeneration of Sitka spruce in west and south-west Scotland (it is also plentiful in the uplands of England, Wales, and Northern Ireland), but considered that it does not provide a practical alternative to planting because seed years are not regular and the seed bearers are prone to windblow. Alternatives to the group system have not been explored in Britain.

Although Gayer (1880) may be said to have systematized group fellings and brought them into prominence, this form of regeneration is of fairly old standing and it was practised here and there before Gayer published 'Der Waldbau'. The group system is essentially one for relatively fertile or well-drained soils and is not suitable for regions exposed to strong winds, or for poor, sandy and otherwise dry soils.

Nowadays the group system is rarely used in its original form, having given place almost everywhere to the strip and group system, which has the advantage of greater security against damage by wind and the extraction of timber (see Chapter 11).

10

The irregular shelterwood system

Fr. Régénération lente par groupe; Ger. Femelschlag

General description

The irregular shelterwood system may be defined as a system of successive regeneration fellings with a long and indefinite regeneration period, producing young crops of somewhat uneven-aged type. The system has several distinctive features. It is a 'natural' system based on characteristics of the indigenous forest types in the regions where it is practised. The object of sustained yield in quantity and quality of timber, whilst maintaining the capability of the sites, is achieved by varying the composition and structure of each stand so as to match differences in site characters. Although use of exotic species is not excluded preference is given to those native to the region.

The irregular shelterwood system draws on elements from several other silvicultural systems, notably the group and selection systems. An important feature is the improvement of the growing stock by selection and tending so that the quality of the timber produced is as high as possible and the capacity for growth of individual trees is fully used. To this end the operations of weeding, cleaning, thinning, increment felling, and regeneration fellings are planned and executed as a continuous process of selection and improvement.

In the irregular shelterwood system close attention is paid to arrangements for harvesting timber efficiently and with the least possible damage to the site and the growing stock. Regeneration and harvesting are closely integrated to achieve a spatial order which brings the stands together into a managed unit.

Improvement and education of stands

Tending begins when the young growth is about 1 m tall. Competing vegetation, climbers, and trees of unwanted species are weeded out and the stocking of the crop is adjusted if necessary. As far as possible mixtures are obtained as separate groups of the different species and if the desired species are lacking they are introduced by planting.

As time passes and weeding gives way to cleanings, a storied structure begins to develop, becoming more apparent as the thicket stage is reached. It is considered that the foundation of the future crop is laid during the thicket stage. The site is crossed with parallel tending trails spaced 20–30 m apart. Cleanings are done to remove trees with crooked stems and large branches, and increase the proportion of well-formed trees. Because of the varying growth in a mixed stand, a dominant upper storey of well-formed trees is gradually formed, together with a vigorous middle storey, and a dominated, mixed lower layer. The function of the last is to suppress branches on the stems of favoured trees and cover the ground.

Selective thinnings begin early in the pole stage. An even distribution of good 'candidate' trees is aimed at. Thinnings are done at regular intervals until the stand reaches an age of 40 to 50 years, by which time the dominants are well-established and the interval between thinnings can be lengthened. In a typical sequence 4000 to 6000 'candidates' are reduced to 800 to 1500 'claimants' and then 200 to 500 'elites' per hectare. The understorey is usually maintained but is sometimes absent. Now follows a final selective thinning to induce 'light increment' on dominant trees of the highest quality. When the increment in value (rather than the increment in volume) of the 'elite' stems in the older age classes begins to fall below an acceptable level, regeneration fellings begin (Leibundgut 1984). The timing of the regeneration fellings and the interval between the seeding, secondary, and final fellings are determined by conditions in the individual stands.

Regeneration fellings

Regeneration begins at certain focal points based on the local terrain and proximity to the main extraction routes. These points are called 'limits of transport'. They are chosen in advance and, as already noted, provide spatial order to the system. However, they need not be regularly spaced; there are fewer on moderate than on more difficult terrain. Where regeneration in groups is to be used, the regeneration fellings begin at one or more 'limits of transport' and proceed outwards towards the extraction routes. As the groups of regeneration are enlarged the seed bearers are felled away from the young trees in the manner described on page 130. The last seed bearers to be felled will usually be close to the extraction routes, and they may be felled in the manner of a strip felling (Fig. 16). Sometimes an extraction route is itself the 'limit of transport'. A gap is formed at the edge of the route and is extended away from it. In time the areas of regeneration coalesce.

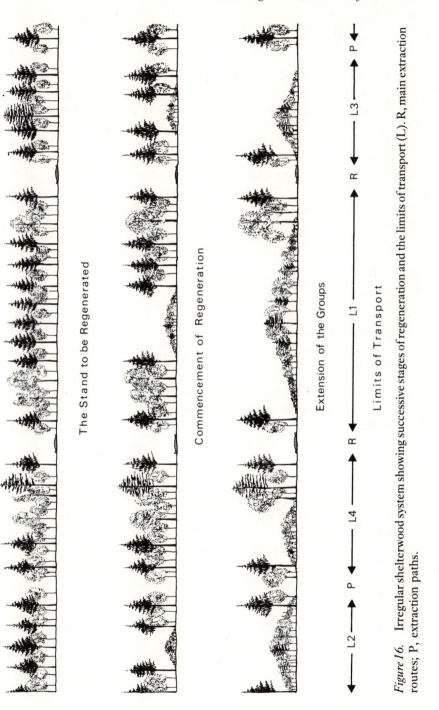

The Stand to be Regenerated

Commencement of Regeneration

Extension of the Groups

Limits of Transport

Figure 16. Irregular shelterwood system showing successive stages of regeneration and the limits of transport (L). R, main extraction routes; P, extraction paths.

Management and yield control

The regeneration period is an indefinite but long one, normally extending up to 50 years; its length may vary somewhat from place to place. Neither the silvicultural system nor the estimation of the prescribed yield require that there should be a definite rotation and usually none is prescribed. There are no periodic blocks; regeneration areas, which may be much scattered, are re-allotted every 10 years, and much latitude is allowed in selecting areas for felling each year and the manner in which the yield is obtained. Close control does exist. The desired growing stock is specified for the whole forest and a felling programme is arranged to provide a sustained yield. Stock maps, aerial photographs, and recurrent inventory are used in keeping control of the growing stock and its increment.

Much that was said about the group system also applies to the irregular shelterwood system, and it is recommended that the two chapters be read in sequence.

Advantages and disadvantages

The chief advantages are:

1. The system is a flexible one and gives scope to a skilled manager.
2. The best possible use can be made of each site.
3. Enhanced increment in volume and value is secured on the best trees.
4. The forest is varied and attractive in appearance.

The disadvantages are:

1. In its most intensive form regeneration and felling are scattered and a dense system of extraction routes is required to offset the effect of scattered working.
2. All the silvicultural operations require skilled people.
3. The system tends to favour shade-bearers against light-demanders unless action is taken to encourage the latter.

Application in practice

There are several variants of the irregular shelterwood system, most of which are found in Germany and Austria. The form developed in Switzerland owes much to the work of Walter Schädelin (born 1873, died 1953)

Figure 17. Irregular shelterwood system. Interior of a stand of beech and Silver fir showing progress of regeneration of beech. The knoll is a suitable limit of transport with access in two directions. Baden Black forest, West Germany.

and Professor Hans Leibundgut (see 'Die Waldpflege' 1984). It is the most recent silvicultural system to have been developed and has replaced all the others except the selection system in Switzerland—where it is used in the mixed forests of the plateau region north of the Alps (Fig. 17).

It seems likely that the system will find limited application in Britain, at least in the form practised in Switzerland. If irregular forests are to expand in Britain the group selection system has already shown its value.

11

Strip systems

Fr. Coupe par bandes; Ger. Saumschlagbetrieb; Span. Cortas por fajas

Strip systems differ in detail but have one common feature, namely that the coupes take the form of quite narrow strips which have certain advantages, mainly of a protective nature, over large coupes. The German word '*saum*' denotes narrow strips, say up to half the height of the old trees or little more, cut along the edge of the strand. '*Streif*' denotes broader strips, as in '*streifenschlag*' (see page 84).

The forms of strip systems described here are:

the shelterwood strip system;
the strip and group system; and
the wedge system.

Two other forms of strip system—progressive strip and alternate strip— are described in Chapter 6.

The shelterwood strip system

Fr. Régénération en lisières obtenu par bandes étroites; Ger. Saumschirm-schlag; Span. Metodo de aclareos sucesivos por fajas

General description

The shelterwood strip system evolved out of the shelterwood uniform system for protective reasons. Under the uniform system the idea of beginning regeneration fellings in the east or north-east and working towards the west or south-west against the prevailing wind direction appeared in some localities about 150 years ago. This procedure required little adaptation to convert it into one of successive regeneration fellings, that is, seeding, secondary, and final fellings made in narrow strips running more or less at right angles to, and advancing progressively against the prevailing wind direction.

Regeneration fellings Regeneration of a compartment begins with a seeding felling of the uniform type carried out along one edge. As soon as regeneration on this strip is sufficiently advanced, a secondary felling is made over it and a seeding felling is done along a second strip adjacent to the first one and on its windward side. When regeneration is sufficiently advanced on this second strip, it is subjected to secondary fellings and the first strip to further secondary and final fellings, and a seeding felling is made along a third strip adjacent to the second one. In this way regeneration fellings advance progressively against the wind direction in a series of narrow strips. The progress of regeneration is represented diagrammatically in Fig. 18.

In the shelterwood strip system, as in the uniform system, the number and frequency of regeneration fellings vary with species, locality, conditions for regeneration, and other factors. In practice the number of successive

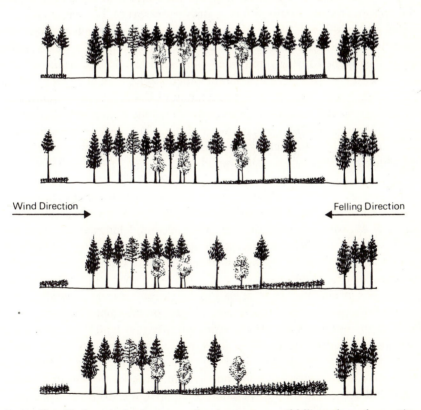

Wind Direction   Felling Direction

Figure 18. Shelterwood strip system showing progress of fellings through a cutting section.

regeneration fellings is not constant over each strip, since regeneration may appear more readily in some places than others, making necessary a more rapid opening of the canopy in the former case than in the latter. Thus the number of successive strips under regeneration at one time may vary, although theoretically there are generally three, in the seeding, secondary, and final stages respectively.

The strip in the seeding stage is generally well-defined, but those in the later stages often merge into each other and their limits are not so readily distinguished. Even after the final felling is completed the regeneration of light-demanders may continue for a time on the cleared strip. As a rule regeneration proceeds more rapidly under the strip system than under the uniform system, due partly to the effect of side light before the seeding felling is made, partly to the larger quantities of seed provided by the adjoining crop, and in some cases to the protective influence of the latter.

The regulation of mixtures is done on lines similar to those adopted for the uniform system. In mixed forests of Norway spruce, European silver fir, Scots pine and beech, the shade-bearing Silver fir and beech are regenerated first by a slight opening of the canopy, Norway spruce appears as the canopy is opened further, and Scots pine seed bearers are retained for a time over the strips from which all the other species have been removed, to regenerate blanks. Sometimes regeneration of shade bearers is stimulated in advance of the regular strip fellings by opening the canopy slightly for some distance ahead of them, or even over the whole area to be regenerated so that when the strip fellings are made regeneration of the more light-demanding species can proceed rapidly.

Form of strips   The breadth of the strips varies according to circumstances. In the Black Forest of Baden-Württemburg, new strips about 20–30 m wide, or occasionally more, are put under seeding fellings at one time; the total width of the belt under regeneration, from the seeding to the final stage averages about 75 m, though the actual width varies according to the progress of regeneration.

On level ground the strips, at least in the early stages, are generally in the form of straight lines and this is also the case on hill slopes in so far as the form of the terrain allows. As the strips advance they tend to become more irregular, because regeneration does not always establish itself evenly throughout their length. The local terrain or the shape of the regeneration areas may require strips that are not straight. In order to present a longer front and accelerate the regeneration of an area, the strips are sometimes given a wavy, serrate, or step-like outline, which has the disadvantage of complicating the alignment of strips.

Influence of wind on the alignment of strips As already noted on page 138, the original idea underlying adoption of the shelterwood strip system in Europe was the need to work against the prevailing westerly wind, so that the newly exposed edges of mature stands should always be protected from it. Hence on fairly level ground the strips are laid out more or less at right angles to the wind direction and advance in a westerly to south-westerly direction against it.

In hilly country, terrain analysis and topex measurements may be used to determine the dangerous local wind direction (see page 32), and the strips are arranged along the contours or up and down the slope so that the wind does not strike against the exposed edges of old stands. Frequently horizontal and vertical strips are both found in the same locality, one or the other being adopted according to wind direction; sometimes both are combined on the same slope with the object of accelerating the regeneration of an area. In the nineteenth century 'felling keys' were prepared as a guide to the alignment of strips and direction of felling in hilly country; that shown in Fig. 19 was prepared for the Neuessing forests in Bavaria in 1885. As stated earlier, models of the terrain can be used to identify where dangerous winds may strike and to plan harvesting arrangements (see page 32).

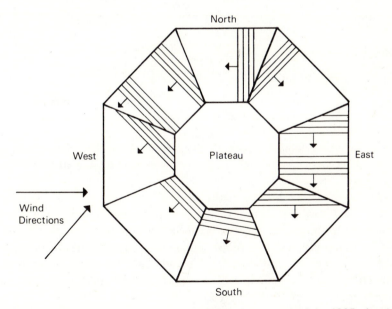

Figure 19. Shelterwood strip system. A felling key prepared in 1885 showing arrangement of strips on slopes of different aspects, when protection against wind is the main consideration.

Influence of lateral shelter on the alignment of strips   There is another group of factors to be considered when deciding on the arrangement of strip fellings—namely those affecting the microclimate of the strip under regeneration. The whole strip can be divided into two parts: the 'inner strip' is the portion opened out for regeneration on which some of the overwood still remains; and the 'outer strip' is where all the overwood has been felled but which receives some side protection from trees standing on the inner strip. The inner strip has overhead and side protection from the sun while the outer strip has side protection only. Under certain circumstances the adverse effect of the sun in drying the surface of the soil and preventing regeneration is so great that strip fellings ought to proceed not from east to west but from north to south, so that the strip under regeneration may receive the lateral shade from the sun which is provided by the old stand to the south of it. When protection against a westerly wind and lateral shelter from the south are both required, this can be obtained by working in eschelon as shown in Fig. 20. This arrangement can also be applied to the strip and group system (page 145).

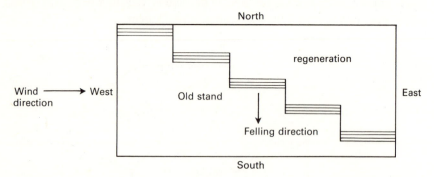

Figure 20.   Shelterwood strip system. Fellings in eschelon to obtain protection against wind and sun. Cutting sections and extraction routes are also indicated in the diagram.

On the other hand, even in the case of Norway spruce, at higher elevations (and northern latitudes) with abundant rainfall and a cool climate, protection from the sun may not be necessary and may even be harmful. It may be found preferable in moist situations to make fellings from south to north, since warmth and abundance of side light are more important than protection from the sun. It will be seen that local conditions of site and climate must always be carefully assessed.

Cutting sections   The rate at which strip fellings advance through an area depends mainly on the speed at which regeneration appears and establishes itself, and therefore varies greatly under different conditions. In the Norway spruce and Silver fir forests of central Europe the average rate of advance can vary from 2 to 10 m each year. Where the rate is 2 m it would take a whole rotation of 120 years to regenerate a compartment 240 m in length, and the crop would consist of parallel even-aged strips of from 1 to 120 years old arranged in regular succession from one end of the compartment to the other. However, in the typical form of shelterwood strip system even-aged crops and comparatively short regeneration periods—sometimes as short as 20 years—are aimed at. Hence the area to be regenerated is divided into a number of cutting sections (see page 67) each of such a length that the strip fellings will advance from one end to the other in the number of years representing the regeneration period. For this purpose it is necessary to fix the regeneration period, if only approximately, and estimate the mean annual rate of advance of the strip fellings. If the regeneration period is taken as 25 years and the average rate of advance as 5 m, the length of the cutting section will be 125 m. If the area to be regenerated is 500 m long it must be divided into four cutting sections. The shorter the cutting sections and the more rapid the advance of the strip fellings, the more even-aged the young crop will be. A slow rate of advance produces a steeper slope in the profile of the young crop than does a rapid rate of advance.

Wide cutting sections, that is, long strips, have two advantages. First, the work is more concentrated and so less complicated and costly than it is where felling and extraction are scattered over numerous short strips. Second, a wide cutting section with no great depth can be regenerated in a shorter time than an equal area of considerable depth but no great width. Therefore regeneration need not be hurried, a matter of importance if supplementary planting is to be avoided.

The formation of cutting sections on level ground is illustrated in Fig. 7, page 68. In forming cutting sections, the risk of exposing unprotected stands to the prevailing wind should be avoided. For this reason, roads, rides, and other permanent lines should be used to separate cutting sections. Where it is necessary to divide the forest into two or more cutting sections, severance cuttings should be made (see page 68).

Advantages and disadvantages

The advantages are:

1. It provides side protection in a definite direction against wind and sun. Side protection is more effective than overhead cover in localities subject to frost and drought.

2. Side light is used to the upmost and overhead light is also available at an early stage, because even sensitive species can be uncovered rapidly owing to the side protection provided by the old stand. This admission of light stimulates development of seedlings and is favourable to the regeneration of light-demanding species.
3. Damage to the young crop during felling and extraction is minimized because trees are felled into the old crop and can be extracted through it.
4. Cultural and tending operations can be supervised easily and the progress of regeneration can be readily followed.
5. Progress of regeneration can also be controlled in relation to the yield. Where it proceeds too rapidly regeneration may be retarded by reducing the number of cutting sections or narrowing the strips, and it may be accelerated by increasing the number of cutting sections or widening the strips.
6. The variations in rates of progress and in sizes of trees often makes this system aesthetically pleasing.

The disadvantages are:

1. The fellings are scattered, but in practice this objection hardly holds, since felling along strips are fairly concentrated and the strips themselves are easily located.
2. The system demands a relatively specialized and rigid layout. Careful design is needed to produce an aesthetically pleasing forest.

Application in practice

The shelterwood strip system was developed from the uniform system more than 150 years ago and has been applied in pure stands of Norway spruce in addition to mixed stands of light-demanding and shade-bearing species. During the first half of this century it gained in popularity, especially in southern Germany where it replaced the uniform system with a short regeneration period. Where reasonably short cutting sections have been formed it has succeeded well. As emphasized earlier, the direction of felling varies with locality and species.

  Between 1898 and 1903 Wagner (1912, 1923) developed a form of the shelterwood strip system which he called 'blendersaumschlag' and Troup (1952, p. 88) interpreted to mean 'fellings aiming at an uneven opening of the canopy along the border strip'. Köstler (1956, p. 284) considered that the whole method failed to win widespread support because it was applied too rigidly, but Wagner's ideas have strongly influenced the development of

European silviculture because he based his method on a thorough study of the factors affecting the success or failure of natural regeneration.

Similar detailed ecological analyses have been the means of solving problems of natural and artificial regeneration in many places. Haufe (1952) reviewed the results of applying the *blendersaumschlag* in Baden-Württemberg during 30 years. He concluded that one of Wagner's chief services was to hasten the abandonment of pure stands and the clear felling system in favour of mixtures of light-demanding and shade-bearing species obtained by natural regeneration.

The strip and group system

Ger. Saumfemelschlag; Span. Cortas por fajas y bosquetes

General description

The strip and group system is a modification of the shelterwood strip system. The general scheme of cutting sections and strips is the same as that of the shelterwood strip system but the fellings are made differently. In the first strip a seeding felling is made consisting of an even opening of the canopy in the usual way, together with freeing of any groups of advance growth in gaps caused by wind and snow. At the same time further groups of advance growth are sought and freed for some distance ahead of the initial strip. If necessary further gaps 30–50 m in diameter are created artificially so as to induce more groups of advance growth to appear. The groups of advance growth are enlarged from time to time as in the group system; together they may comprise 20 per cent of the area being regenerated. Meanwhile the strip fellings advance steadily, absorbing the groups of advance growth and new clearings are also pushed on some distance ahead of the strip fellings until the end of the cutting section is reached.

Figure 21 shows the progress of fellings under the strip and group system. In some cases group fellings are made all over a cutting section before strip fellings are begun and the latter are largely in the nature of removal fellings. The progress of the fellings and the form of the strips are less regular under this system than under the shelterwood strip system.

The strip and group system is well adapted for regulating mixtures, on the lines already indicated for the group system (page 128). Shade-bearers may be introduced artificially in small gaps ahead of the strip fellings; beech is frequently introduced in this way into conifer forests, the conifers being afterwards regenerated naturally by strip fellings. Seed bearers of

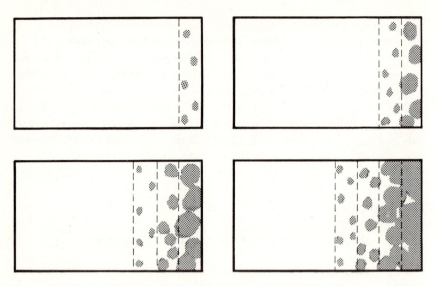

Figure 21. Strip and group system showing progress of felling. Regenerated areas are indicated by the shading.

light-demanding species are generally retained on the strips until the final felling, so as to seed up blanks; such species may also be introduced artificially on the strips when they have been opened out.

Advantages and disadvantages

The advantages are:

1. Regeneration can be accomplished more readily and rapidly by establishing ample groups of advance growth ahead of the strip fellings.
2. The uneven-aged form of the young crop gives some protection against snow and wind.
3. Mixtures can be regulated relatively easily.
4. The appearance of the forest is aesthetically attractive.

The main disadvantage is risk of damage to groups of advance growth when extracting logs, particularly on steep slopes.

Application in practice

The strip and group system was developed chiefly in Bavaria by H. von Huber, chief of the Bavarian Forest Service. It is widely used in central Europe.

The wedge system

Fr. Régénération par coupes progressives en coin; Ger. Keilschirmschlag; Span. Corta de abrigo en cuña

General description

The wedge system or shelter wedge system is a form of strip system in which the fellings, instead of starting at one end of a cutting section and proceeding to the other, begin by a strip in the centre of it and proceed outwards in both directions (Fig. 22).

On level ground or gentle slopes the wedge fellings are begun by laying out parallel lines running from east to west or from north-east to south-west in the direction of the prevailing wind. These lines are spaced at intervals of 80 m through the compartment to be regenerated. They are felled to a width of 2–5 m and form the initial strips. Protective belts of forest are left standing at each end of the felled strips.

Midway between the initial strips and parallel to them, extraction racks are laid out leading to the main routes which bound the compartment. Subsequent fellings widen the initial strips by progressive narrow strip fellings made at frequent intervals, not parallel to the initial strips but at a slight angle to them so as to form gradually widening wedges with their apexes pointing towards the prevailing wind direction. At the same time the apexes

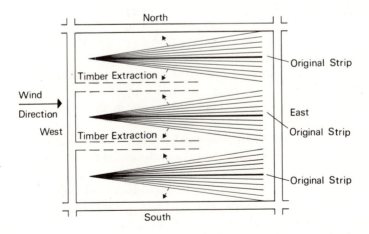

Figure 22. Wedge system on level ground showing original strips and subsequent enlargement.

Figure 23. Wedge system. Interior of mixed stand looking along a ride running parallel to the axis of the wedge. Villingen-Schwenningen, Black forest, West Germany.

are pushed forward towards the edge of the regeneration area. Seed bearers of the light-demanding species, such as Scots pine and European larch, are left for a time to regenerate blanks as the wedges are opened out. Eventually the wedges join up and the protective belts are removed by strip fellings, so that the whole area becomes regenerated. The last trees to be felled are those along the sides of the main extraction routes; a few of these, usually Scots pine, are left to put on increment and act as seed bearers for the regeneration of blanks. Different stages in the progress of wedge fellings are shown in Figs 23 and 24.

Great stress is laid on measures which promote and complete regeneration. These include thinnings throughout the life of the stand increasing in frequency in the later stages until the regeneration of shade-bearers takes place; removal of raw humus or cultivation of the soil, where needed, to stimulate natural regeneration; and rapid removal of the old crop as soon as regeneration appears over the area. Thus the wedge fellings are largely removal fellings over an area already partly regenerated and the more rapidly the trees are removed the better. On the plains of the Rhine, in

Figure 24.   Wedge system showing regenerated crop with reserves of Scots pine. Villingen-Schwenningen, Black forest, West Germany.

forests of Scots pine mixed with broad-leaves, whole compartments are regenerated in 6 to 8 years; in the Black forest, where Norway spruce and Silver fir predominate, the period is usually longer.

It is claimed that wedge fellings tend to reduce the risk of damage by wind. In the plan adopted on level ground the wind enters the narrow end of the wedge and its force is dissipated as it passes through it. On westerly slopes the same happens. On easterly and north-easterly slopes the point of the wedge has to be opened out so that there is a clear lane from top to bottom of the slope that permits winds to pass through. On northerly and southerly slopes the wind channel formed by widening the wedges serves the same purpose. Philipp (1926) constructed a special felling key for the wedge system based on a study of wind on different aspects in the forests of Huchenfeld in the Baden Black Forest. Where wind is not a source of damage, alignment of the wedges is governed by the shape of the area and the position of roads.

On steep slopes the procedure is modified for extracting logs downhill.

The initial strips are cut vertically up and down the slope and the wedges are widened in such a way that their apexes point downhill. Fellings begin in the east of the area and continue successively towards the west, so as to retain protection from the west wind for as long as possible. After the wedges join up, a gradually widening channel for the wind is formed along the bases on the upper part of the slope.

Advantages and disadvantages

The advantages are:

1. The long front presented for regeneration enables the old crop to be removed rapidly.
2. Timber can be harvested cheaply, methodically, and with little or no damage to the young crop.
3. Considerable reduction in damage by wind has been claimed.

The disadvantages are:

1. It is quite complicated to apply in practice.
2. The northern side of the wedge is exposed to the sun. In dry situations this might have an adverse effect on natural regeneration.

Application in practice

The system originated in Baden-Württemburg in West Germany during the first quarter of this century. The proposal to make cuttings in wedges pointing towards the prevailing wind direction so as to reduce the risk of damage by wind was made by Eifert (1903) when describing the results of his study of wind in the north-eastern part of the Black Forest. Eberhard (1922) put the proposal into practice in the forests of Langenbrand in the same region, with promising results. Philipp (1926) elaborated and systematized wedge fellings in the forests of Huchenfeld in the Baden Black Forest and, later, as chief of the Baden Forest Service encouraged their wider use.

The forests of Langenbrand and Huchenfeld consist chiefly of Silver fir, Norway spruce, and Scots pine, situated in hilly country where the climate is somewhat cold and rough. The annual rainfall lies between 890 and 1020 mm and gales from the west and south-west are a constant source of danger. In the first instance, one of the chief objects of wedge fellings was to reduce the risk of wind damage. Subsequently, however, the wedge system was introduced into forests of Scots pine mixed with hornbeam, beech, oak, and other broad-leaves on the plains of the Rhine, where the danger from storms is far less serious.

Returning to higher ground, the forest district of Villigen-Schwenningen lies on the east side of the Black Forest in Baden-Württemburg. The forest is composed of Norway spruce (62 per cent), Silver fir (18 per cent), and Scots pine (20 per cent), and the yield classes are respectively 8, 10, and 6. The soils are podsols derived from sandstone of Permian age; they are low in nutrients. Frosts occur in every month and damage to the trees is common. The clear cutting system was abandoned in 1950 and the wedge system substituted so as to reduce wind damage and provide protection for the regeneration. In addition an increased proportion of Silver fir, to 30 per cent of the crop, is desired because of their deep rooting.

The size of the regeneration area is 30 ha and the wedges are aligned in a south-west direction. The old crop is repeatedly thinned until Silver fir appears; the thinning is then intensified to encourage Norway spruce, and a new strip is cut every 5 years to widen and lengthen the wedges and provide overhead light for Scots pine. Some Scots pine seed bearers are reserved for one further rotation to fill blanks and put on light increment.

The extent of the wedge system in use is only 20 to 30 thousand ha and most of that area appears to be on podsols. Nevertheless it may be useful in solving some of the problems of temperate forestry in windy climates.

12

The tropical shelterwood system

Fr. Coupes progressives sons les forêts tropiques; Ger. Tropisches schirm-schlag

The advantages of concentrated working and regeneration, as compared with fellings scattered far and wide over tracts of tropical moist forest, are so great that several attempts have been made to develop systems of regeneration based on the principles of the uniform system. The name given to these is the tropical shelterwood system.

General description

The tropical shelterwood system is an adaptation of the uniform system to tropical moist forest in which the seeding felling consists of a general opening of the canopy by cutting climbers and progressively reducing the middle tree storey by felling, girdling, or poisoning the unwanted trees (Robertson 1971). In converting from natural indigenous forest to more uniform high forest, the fellings recognized are generally of two kinds:

regeneration fellings; and
selection and improvement fellings, including thinnings.

Regeneration fellings can seldom be done with the precision character-istic of the uniform system itself. In many cases they consist of freeing re-generation that is already present, while supplementary artificial regenera-tion may also be needed. The resulting young crops are often somewhat irregular, but they can also be quite regular (Wyatt-Smith 1987). Some-times undisturbed tropical moist forest contains much young growth and here the fellings consist of freeing the regeneration from overhead cover, after the manner of secondary fellings.

Selection and improvement fellings are intended to utilize mature timber (and for this reason are sometimes called 'salvage' fellings) and at the same time remove defective or otherwise undesirable trees interfering with more promising members of the crop. These fellings pass over those portions of the forest that are being prepared for regeneration (see page 228).

The Malayan uniform system

The canopy of lowland evergreen dipterocarp rain forest in West Malaysia has a well developed, multi-storied structure which has been fully described by Whitmore (1984). The emergent layer comprises trees of large size that have their crowns free, above the closed middle and lower layers. Below these are a layer of saplings and young trees and the herbs and tree seedlings of the forest floor. These forests are composed of mixtures of many species varying in their silvicultural characteristics. The commercial value of the timber produced by the different species also varies greatly.

In the 1930s, silviculturalists in Malaysia began to explore ways of regenerating the forests naturally while simplifying its structure and increasing the proportion of principal species in the new crop. The shelterwood uniform system appeared to have several advantages as a means of obtaining natural regeneration. It was found (Barnard 1955) that in virgin forest seedlings of the principal species were almost always present in sufficient numbers to give prospect of successful restocking. Although these seedlings were often suppressed and their growth was very slow, they responded almost immediately to increased light (see page 229). It appeared that conditions on the floor of virgin forest are favourable for the germination of species of the *Dipterocarpaceae* (which must germinate immediately or lose their viability) and that seedlings of many species are able to survive a period of suppression.

The next step was to ensure the survival and encourage the growth of seedlings of the principal species. Under natural conditions a light-demanding, high forest species can only grow rapidly where a gap has been created in the canopy by a fallen or dead standing tree; such an opening is usually short-lived. Seedlings are stimulated to grow by the overhead light but so also is the undergrowth and the crowns of saplings, poles, and trees in the various canopy layers surrounding the opening, so the gap is soon closed. The first attempts to create artificial openings were made by girdling the stems of trees, but the action of girdling was slow and erratic. A great advance was made by irrigating the girdle with a herbicide. It then became possible to time the opening of the canopy relatively accurately.

The ground under a large artificial opening receives direct overhead light for several years but other parts of the canopy remain closed and any natural regeneration there stagnates. It was found that conditions favouring rapid height growth of the principal species had to be made uniform over whole compartments, and this could be done either by removing the existing canopy completely, leaving immature trees and advance growth, or removing the existing canopy progressively by several poison-girdlings plus fellings of the principal and some secondary species.

The admission of light to the forest floor enabled seedlings of the primary species to grow but also encouraged rapid growth of secondary and weed species, climbers, and ephemeral herbaceous species which, after 1 or 2 years, threatened to suppress and kill the principal species. At first repeated cleanings were done to free and favour them, but about 1937 (Barnard 1955) some experimental plots were left untreated and it became evident that the faster-growing principal species eventually gained dominance unaided, leaving a new lower storey to re-form below them. Provided there is abundant growth of erect trees and the tangle is left alone, the period when climbers are rampant is short and the trees eventually win through. Persistent climbers were cut and dominant saplings of the primary species were released from competition by selective girdling and poisoning of their neighbours. It was found essential to allow a closed or nearly closed canopy to form and resist the temptation to assist saplings below its level. A valuable new crop was assured if good stems of the principal species occur at 100/ha. The later cleanings were done stem to stem, that is, wherever regeneration of the principal species was absent, saplings of other species were retained usually at a spacing of 2 m between stems. The early formation of a canopy was considered desirable to control climbers and conserve soil fertility.

The Malayan Shelterwood system was systematized by Wyatt-Smith (1963) who formulated the results of 20 years of experiment into detailed silvicultural prescriptions. He laid great stress on diagnostic sampling of regeneration before the principal felling to decide whether it should be permitted; and after felling to assess the need for tending treatments to ensure that the regeneration was adequate in quantity, species composition, and condition to satisfy the objects of management.

The key to the system is the presence in the forest at the time of felling of well-distributed seedlings of the desired species (Schmidt 1987) particularly *Shorea leprosula* and *S. parvifolia*. These species are fast-growing and produce timbers of the light red Meranti group. They fruit gregariously and seedling populations are replenished every 2 or 3 years. Progressive mortality among the seedlings follows and just before a fruiting the seedling stock may be inadequate in amount and distribution to ensure a dense new crop. The system also allows light-demanders other than dipterocarps to succeed, notably *Dyera costulata*, *Endospermum malaccense*, and *Pentaspadon motleyi*. Felling kills many seedlings and saplings but releases the rest.

Application in Uganda

Budongo forest

In Uganda the tropical semi-deciduous forests occur in two main groups. The first is a discontinuous belt 50–80 km wide forming a crescent around the northern and north-western shores of Lake Victoria. These are known as the lake-shore forests and one of them, called South Mengo, will be discussed later. The second, and larger group of forests, lies on the uplifted eastern rim of the rift valley above Lakes George and Albert. They extend from Masindi in the north to Rwanda in the south. These are called the western forests and they include Budongo, Bugoma, and Toro.

Budongo forest lies in the kingdom of Bunyoro at elevations of 900–1000 m. It extends to nearly 500 km^2. The rainfall is 1400 to 1500 mm and it falls in two periods, from the end of March to May and from August to November.

The forest has a fully developed multi-storied structure. The greater area consists of mixed *Khaya anthotheca*, *Entandrophragma angolense*, *E. cylindricum*, *E. utile*, *Lovoa brownii*, and *Chlorophora excelsa*. The mahoganies of the *Meliaceae* are particularly valuable, but some of the species of the colonizing stage, such as *Maesopsis eminii* are also important commercially. This forest type is gradually being invaded by the gregarious *Cynometra alexandri*.

In 1933 the forester and ecologist W. J. Eggeling was posted to Budongo forest. He set out permanent research plots in the forest types that he recognized, and gradually worked out methods for converting Budongo to sustained yield management whilst conserving its characteristic plant and animal species.

The prescriptions of the first working plan were followed from 1935 to 1944. It provided for exploitation of the forest subject to its perpetuation, and this object was reinforced by the first revision which remained in force from 1945 to 1954. This prescribed for systematic removal, during 40 years, of very large trees producing primary timbers that were unlikely to remain standing whilst the forest was being converted to sustained yield working. When these salvage fellings began only the mahoganies, *Maesopsis eminii* and a few other species with particularly durable timbers such as *Mildbraediodendron excelsum* and *Erythrophleum suaveolens* were marketable. The felling limit for the mahoganies was 80 cm diameter and 60 cm for the other species; the annual yield was controlled by the volume of sound logs.

The regeneration block was enriched by planting striplings in lines (see page 230). The surge of climbers was controlled by cutting them and trees of undesirable species were also cut to release the planted mahoganies. Despite the speedy growth of the competing vegetation, attacks by shoot borers (*Hypsipyla* spp.) and Gall fly (*Phytolyma lata*), browsing by antelope, and damage by elephants, the enrichment was often successful because the mahoganies benefited from the release from competition and overhead shade.

The working plan was again revised in 1955. By this time more species had become marketable, the prescribed yield had been increased, and the intensity of the salvage fellings rose leaving the canopy more broken. Regeneration of desirable species was found in almost all the gaps and the possibility arose of restocking by natural regeneration.

The forest was divided into compartments of 300 ha so that silvicultural operations could be planned, executed, and controlled. The road network was improved to provide access for vehicles carrying men, equipment, and materials. Two years before the salvage felling was due, the access road to the compartment was made in consultation with the concessionaires, who constructed them. During the next 12 months access paths were cut which divided the compartment into units of 4 ha. The marketable trees that had to be felled were completely enumerated. Climbers were cut and herbicide applied to undesired weed species and to trees with defective stems. The salvage felling was then done.

Ten years later, when the canopy of pioneer species had raised the tangle of climbers sufficiently to allow entry, diagnostic sampling was done to establish the extent and condition of the regeneration. If necessary the saplings were released from competition by cutting climbers and applying herbicide to relict trees that had not died and to other trees that were now suppressing the regeneration.

In the mid 1950s Dawkins (1959) became concerned about the damage being done to advance growth when large trees with heavy crowns were felled. The third revision of the working plan (which was approved in 1965) therefore provided for conversion of the forest, after the salvage felling, to a single cycle of felling and regeneration. The converted forest would consist of many species with a large range of diameters due to the retention of advance growth at the time of conversion. It would show great variation in rate of growth even among individuals of the same size and species, and would contain regeneration of species that naturally occupied different levels in the canopy.

The salvage felling and the conversion cycle now proceeded simultaneously but in different parts of the forest. The latter followed 40 years after the former.

As the area of forest that was being 'improved' rose the numbers of animals feeding in it also rose. Colobus monkeys (*Colobus* spp.), chimpanzees (*Pan troglodytes verys*), and birds such as the forest lorey (Psittacidae) and

Guinea fowl (*Numida meleagris*) liked feeding in the treated forest—especially in areas of the *Cynometra* association. For various reasons, not all connected with the management of the forest, numbers of elephants also rose and the damage done in the regeneration areas became serious. Access tracks were kept open to allow the forest guards to patrol and chase out the herds as soon as possible. By the late 1970s illegal shooting had reduced the numbers of elephants and the problem in Budongo abated.

Natural regeneration has been used to perpetuate the principal species in the mixed semi-deciduous forests of western Uganda for 30 years and, provided that the new crops are not suppressed by climbers or an overwood, success is likely. An average diameter increment of 0.5 cm/year on 50 stems/ha is feasible (Philip 1986).

South Mengo forest

The method of regeneration developed in South Mengo forest depended for its success on several interrelated factors. The lake shore forests in Uganda are situated near to urban areas so the demand for sawn timber and fuel is high. New wood-using industries have been built so many more species became marketable. Natural regeneration of the principal species was shown to be more common than had been supposed; and methods for enrichment by planting the fast-growing *Maesopsis eminii* and other colonizing species had been devised.

In the regeneration block, felling coupes of 40 to 80 ha are allocated to concessionaires who fell all well-formed trees of the marketable species. Trained staff of the Forestry department then select and mark young trees of the principal species to form part of the next crop. They also fell trees of weed species and defective trees of the marketable species with stems of greater than 60 cm diameter, so as to avoid damage to the marked stems. Charcoal burners and firewood merchants are then allowed to fell and convert all other unmarked trees and the slash. One coupe has to be prepared satisfactorily before another coupe can be opened.

Immediately after this preparation of the site, species such as *Maesopsis eminii, Terminalia ivorensis, T. superba,* and *Cedrela odorata* are planted in gaps—either singly or in pairs. These grow up to 3 m each year on sites temporarily sterilized by the traditional earth kilns where weeds are slow to become re-established.

The advantages of this method are:

1. Firewood and charcoal are made available and the utilization of the existing forest is greatly improved.

2. The growing stock can be enriched with desirable species.
3. The new crop develops in conditions encouraging rapid growth and establishment.
4. The total cost of regeneration is reduced.
5. The land is seen to be more intensively used and in consequence there is less temptation to encroach on it to grow agricultural crops.

Application in India

Shorea robusta or sal is a gregarious tree which grows over a wide range of soils and climates in India. Its regeneration presents many problems that have not yet been completely solved. The following account refers almost entirely to the moist, high-level alluvial forest in Uttar Pradesh, in which the sal is of especially fine quality and has therefore long attracted attention.

The fruiting of sal is periodic; in years when there is a combination of abundant seed with the onset of the monsoon rains at about the time of ripening, seedlings appear in great abundance in the thin grassy floor of frequently burned woodland. If the rains are late, the seed will have lost its viability by the time that they come, while if they are early, the gales that accompany their onset may blow down the seed before it is ripe; hence years when seedlings are abundant are rare and come at irregular intervals. At first, the seedling is easily killed by fire, but after it is 1 or 2 years old the shoots may be burned repeatedly for many years yet fresh shoots of gradually increasing vigour are produced from the rootstock. Moreover, seedlings are moderately light-demanding and intolerant of competition, and will eventually succumb in the struggle with dense grass, or with the broad-leaved undergrowth that forms in the absence of burning. The dense weeds that come after too heavy felling are equally liable to suppress seedlings that are not sufficiently vigorous. Therefore a dense canopy and also a very open one both prevent establishment. Grazing of the young tender shoots that come up immediately after burning is also very injurious.

It has been usual to select for regeneration periodic blocks in which abundant advance growth was already present. In some districts it is best to clear fell the overwood; in Uttar Pradesh, however, a shelterwood must be left both as a protection against frost and in order to restrain rapidly growing weeds that are more light-demanding than the sal. After a felling which leaves 50 to 75 trees/ha, the ground is burned just before the rains and deer, which are often abundant, are excluded. Subsequent weeding is essential, and is far more effective if done during the moist growing season than during the (dry) resting season, but it is often very difficult to obtain labour during the rains. After 5 or 6 years, thinning of the regeneration and gradual removal of the overwood may begin, the process being completed in 10 to 12 years.

Building up a stock of suppressed advance growth is far more difficult and uncertain than the release of existing advance growth. The exact condition of the canopy and the ground cover are apparently very important. A moderate opening of the canopy, preferably by making small gaps and retaining a mixture of species and some middle storey (for sal establishes itself best beneath species other than itself), together with some burning and weeding during the growing season is believed to be the best treatment. Opinions have varied considerably on how much fire should be used, and there is no doubt that it varies with the locality. The aim is to get a thin cover of mixed grass and *Clerodendron* in approximately equal amounts; too much weeding, too much burning, or too heavy an opening of the canopy can all produce a dense sward of grass in which seedling establishment is impossible. Scarification is beneficial and may now be possible. Present indications are that building up sufficient advance growth takes at least 6 or 7 years and may take 20 to 30 years.

Thinning tropical moist forest

In the shelterwood uniform system, the life of a stand in a forest managed for sustained yield is divided into two parts, those of 'education' or preparation for regeneration, and regeneration itself. There are several features of tropical moist forest that pose problems for silviculturists and forest managers when converting them into forests managed for sustained yield. According to Philip (1986) the most important features are (Fig. 25):

(1) the large number of species per unit of area and their diverse ecological status;
(2) the layered or storied nature of the forest, each storey containing species whose height at maturity is limited, as well as species typical of the storeys above;
(3) the good stem form and timber quality of many of the trees in the highest storey, including the emergents which may attain heights of 40–45 m;
(4) the frequency of buttresses which hamper measurement and tending;
(5) the presence of strangling figs (*Ficus* spp.) and epiphytes;
(6) the abundance of herbaceous and woody climbers or lianes;

Figure 25.   A simplified plan and elevation of exploited tropical moist forest, showing the layers or storeys, emergents with large crowns and tall stems, buttressing, and woody climbers. Based on Jones (1948).

(7)   the complex spatial pattern of species, which often resembles a mosaic of associations superimposed on a broader pattern produced by topography and drainage.

One object of silviculture and management for timber production is to reduce the complexity of the forest and increase the proportion of principal species. The purpose of this section is to describe an experimental thinning of moist tropical forest in West Africa, which could eventually lead to successful regeneration and sustained yield working.

In 1976 the Societé pour la developpement des plantations forestiers (SODEFOR) with technical assistance from the Centre Technique des Forêts Tropiques established experiments in the Republic of Ivory Coast on the west coast of Africa, to study the response of dense tropical forest to different degrees of thinning. The trials, each covering 400 ha in 900 ha of undisturbed forest, were laid out on sites in three ecological zones, namely tropical semi-deciduous forest (at La Téné), tropical evergreen forest (at Irobo), and forest of intermediate type (at Mopri). The principal species include *Aningeria robusta*, *Gambeya delevoyi*, *Guarea cedrata*, *Khaya anthotheca*, *Nesogordonia papaverifera*, *Scottellia* spp., *Tarrieta utilis*, and *Triplochiton scleroxylon*. Data are being collected for 42 principal species; many are present on two of the sites but only few are growing at all three sites.

The specific objects of the trials are to compare the effects of two weights of thinning, mainly of secondary species, and one controlled exploitation type of felling of principal species on:

(1) the growth and yield of each principal species in the different stands;
(2) the response of the stands as a whole in terms of induced mortality, recruitment of young stems, and development of climbers and coppice shoots;
(3) the occurrence and growth of natural seedlings and saplings of the principal species.

The total basal area of the undisturbed forest was 27.9 m^2/ha at La Téné; 24.5 m^2/ha at Irobo; and 22.6 m^2/ha at Mopri. The thinning treatments (done by girdling plus herbicide) removed 30 or 40 per cent of the total basal area, beginning with the largest trees of the secondary species. There are untreated control plots. The controlled exploitation felling of principal species was done at La Téné. Trees with stem diameters of 80 cm or more were elled and extracted by skidding. The average standing volume of the undisturbed forest at La Téné was 100 to 150 m^3/ha for the principal species and 270 m^3/ha for all species. The volume removed was 53 m^3/ha.

At each site the treatment plots were 16 ha in extent; the thinning treatments were applied to 9 ha; and assessments are being made on plots of 4 ha each. The positions of all trees of the principal species 10 cm or more in diameter were plotted and diameter increment has been monitored at 2-year intervals, together with deaths. Counts of seedlings and saplings were made within the assessment plots so that the progress of natural regeneration could be followed.

After 6 years some results were reported by Maitre (1987):

1. The girdled and poisoned trees soon disappeared and the overwood became more open, consisting of sound trees without climbers or drooping crowns.
2. The individual species, both principal and secondary, have responded in diameter growth to the removal of mainly secondary species.
3. The thinning had not altered appreciably the composition of the understorey of seedlings, saplings, and small trees. Climbers and coppice shoots had not hampered the regeneration.

4. Following the controlled exploitation felling the gains in diameter growth and recruitment of young trees were consistently below those recorded after thinning. In addition large gaps were left in the canopy at erratic intervals.

The figures given by Maitre (1987) for annual increment per cent of the principal species were:

Untreated control plots   0.5–2.0 per cent
Thinned plots   2.5–3.5 per cent
Controlled exploitation felling 1.5 per cent

There is widespread concern about the present condition and future prospects for tropical moist forests, the plants and animals associated with them, and for the indigenous people who depend on these forests for their livelihood. Wyatt-Smith (1987) drew on his wide experience of silviculture and management in tropical moist forests to state these principles:

1. Each country with tropical moist forest must create a permanent forest estate as part of an integrated policy for the use of land.
2. Each country must also form a strong forest service, manned by professionals who are given control over the forests they are required to manage. They must have clear objects of management to guide their work.
3. More information must be accumulated by research about the many different ecosystems in tropical moist forests so that appropriate silvicultural systems and management can be applied to sustain the yield of timber and conserve these ecosystems.

13

The selection system

Fr. Jardinage par pieds d'arbres; Ger. Plenterhieb, plenterung; Span. Corta por entresaca, metodos de selección

General description

The selection system differs from all the other silvicultural systems in that felling and regeneration are not confined to certain parts of the forest but are distributed all over it, the fellings removing single trees or small groups of trees selected throughout the forest. Fellings done in this manner are termed selection fellings; they result in an uneven-aged or irregular type of forest in which all the age or size classes are mixed together over every part of the area. This completely uneven-aged type of forest is usually termed 'selection forest' or 'forest of selection type', whether it has actually been produced by selection fellings or not.

Felling and regeneration

Under the selection system, scattered single trees or small groups of trees are selected over the whole area and removed. Where conditions are favourable, natural regeneration springs up in the gaps so created. Under ideal conditions this process goes on year after year over the whole forest, the volume removed being fixed by rules of management (page 54). This results in the constant maintenance throughout the whole area of an uneven-aged or irregular structure in which trees of all ages are mixed together. Such perfect distribution is seldom actually found, the age-classes occurring more usually in small groups resulting from groups of regeneration springing up in gaps. The appearance of forest worked under the selection system is shown in Figs. 26 and 27.

In their most primitive form, selection fellings consist of removing all trees that have attained a certain diameter, sometimes with the proviso that any trees required as seed-bearers should be retained. Such crude and unregulated selection fellings are mere exploitation requiring little or no

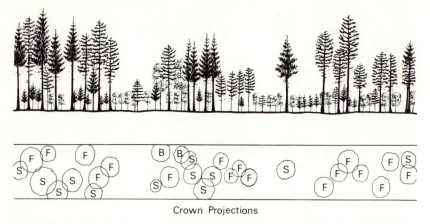

Crown Projections

Figure 26. Selection system. Side elevation and plan of forest of selection type containing three species. S, spruce; F, Silver fir; B, beech.

silvicultural skill; they do not ensure regeneration nor do they provide for a regular sustained yield and they often result in serious deterioration of the forest.

If selection fellings are to be raised to the status of a silvicultural system, something more is required than mechanical removal of exploitable trees. It is necessary also to provide for sustained yields by making thinnings among the various age classes to ensure:

(1)   that these are maintained in their correct proportions;
(2)   that a suitable mixture of species is maintained, if this is necessary;
(3)   that young saplings are freed from suppression; and
(4)   that defective stems of any size are removed wherever they are hampering better ones.

Hence, wherever the selection system is worked as a true system, thinnings among the immature stems are made along with fellings of trees of exploitable size, the whole forming one operation. Cleanings are also done amongst the younger age classes and measures to assist regeneration may include removal of raw humus, soil cultivation, and even direct seeding or planting in gaps caused by felling or otherwise.

Fellings under the true selection system generally involve removal of:

(1)   dead and dying trees;
(2)   trees that are diseased, misshapen, or otherwise defective or lacking in vigour or of an undesirable species, particularly if they are interfering with better stems or promising groups of young growth; and
(3)   trees of exploitable size, particularly if defective or lacking in vigour.

Figure 27. Selection system. Interior of stand showing intimate mixture of species and age classes (Ammon 1951).

Nowadays, the tendency is to abandon the idea of an exploitable diameter and leave especially vigorous, well-formed trees of any size to put on increment.

The felling cycle

Fr. Rotation; Ger. Hiebsumlauf; Span. Rotación periódica

The selection and felling each year of scattered trees over the whole forest would be impracticable except in areas of small size, and hence this procedure (which is sometimes called the 'ideal' selection system) is usually confined to small forests belonging to private persons or communities. In forests of larger size it is customary to divide the area into several more or less equal blocks, in one of which selection fellings are made each year, so that the whole forest is worked over during a period of years, called the felling cycle, which is equal to the number of blocks. Under this 'periodic' selection system, fellings are more concentrated, there being an accumulation of mature trees during the interval between two successive visits to the same block, so that a larger volume per hectare is available over the annual coupe than in the case of annual fellings extending over the whole forest.

The felling cycle is fixed according to the circumstances of each case. If it is very short, involving large coupes, the cultural advantages of the selection system—including cleanings and thinnings—have full play, but the work is scattered and therefore costly. If it is very long, involving small coupes, fellings over definite coupes become intensive, larger quantities being removed over a restricted area; this is an advantage from the economic point of view and it may also favour the regeneration of light-demanders through the creation of large gaps, but the cultural advantages are to a great extent lost, while the accumulation of large trees upsets the correct distribution of size classes and destroys the character of true selection forest.

In Europe, the felling cycle is generally not more than 10 years and is often less. In France, it is usually between 5 and 8 years, seldom if ever more than 8 years and in Switzerland 6 to 10 years. A definite felling cycle is sometimes dispensed with in coniferous forests in regions where frequent windfalls make necessary quick removal of fallen trees in various parts of the forest; under such conditions a pre-arranged felling cycle with definite annual felling areas would be inadvisable. This procedure would be difficult, if not impossible, over extensive forests where close personal attention could not be exercised.

The structure of the growing stock

In each part of a true selection forest with a mixed and stratified structure, there is a continuous series of age classes, and continual recruitment to the growing stock by natural regeneration. The distribution of stem diameters is such that each diameter class has fewer stems than the adjoining smaller diameter class, and also that the ratio of the number of stems in a class to the number of stems in the adjoining class is constant. De Liocourt (1898) was the first to derive a model for this type of distribution in selection forests and his negative exponential model is:

$$Y = ke^{-aX}$$

where Y is the number of stems/ha, X is stem diameter at breast height, k is a constant reflecting the stocking of very small seedlings, and a is a constant governing the relative frequencies of successive diameter classes. The constants k and a vary with species and sites; they are calculated for each selection forest using data from inventories of the growing stock.

The negative exponential model can be used to estimate the normal distribution of diameter classes in a selection forest and selection fellings are done so as to lead gradually toward the 'normal' condition. Comparison of the actual distribution of diameter classes with the 'normal' serves as a guide, the fellings being confined as far as possible to those classes that are in excess (see Fig. 28).

Regulating yield

In Switzerland, the regulation of the yield of timber from selection forests is by the *Méthode du Controle* or check method devised by Biolley (1920). He adopted the principle of gaining maximum increment from the smallest possible growing stock, and he set out to maintain a definite distribution of size classes so as to establish a normal forest and ensure sustained yields. Biolley divided the forest into compartments and made recurrent inventories of the growing stock in three size classes (large, medium, and small) at intervals of a few years (usually 6 to 10 years) to determine the relation between increment and growing stock, fix the yield for the next period, and plan the fellings so as to work toward the normal distribution of size classes. The check method is described in detail by Knüchel (1953) and in outline by Osmaston (1968).

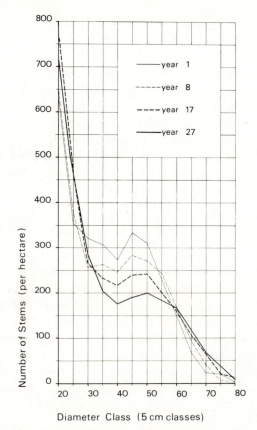

Figure 28. Selection system. Changes in the stem number curve in a managed selection forest during 27 years. Based on Knüchel (1953, p. 82).

Advantages and disadvantages

On the understanding that the term 'selection system' refers to the system in its true form, with selection fellings made to improve the structure of the stands, the quality of the trees, and hence the value of the crops—and not to the mechanical removal of trees that have attained a certain diameter, the chief advantages and disadvantages may be stated as follows.

The advantages are:

1. By maintaining a constant forest cover, exposure of the soil is reduced and protection provided against erosion, landslips, and avalanches.
2. Damage by wind and snow is minimized and disorganizes working but little (see pages 15 and 36).

3. All seed years can be used for regeneration and seedlings are well-protected.
4. The system is very flexible and intensive so that best possible use is made of each part of the site, and the productive capacity of the soil can be conserved.
5. There is no need to maintain as high a proportion of young trees to old as in regular forest where there must be equal areas of each age class. Hence a larger proportion of the growing stock and yield can be in the form of trees of large size and high value.
6. It is possible to promote the development of individual trees with good form and branching habit and retain them as long as they are making valuable increment. A tree need not be felled merely because it has reached a certain age. Because the crowns of the dominant trees are well-developed and stand more or less isolated, increment continues undiminished to an advanced age.
7. The lower parts of the stems of trees of large size yield timber of high quality. But the upper parts of the stems do bear large branches, producing large knots, and taper markedly.
8. Aesthetically, selection forest is often considered attractive and this can bring benefits to the owner, the region, and the general public.

The disadvantages are:

1. Considerable silvicultural skill is needed in marking fellings. It is one of the most intensive systems.
2. Felling and extraction must be done with skill and care.
3. The system is applicable chiefly to shade-bearers. Light-demanders may be treated by group selection methods (see Chapter 14).

Several aspects of the selection system have long been debated. It is often said that owing to the scattered nature of the fellings and regeneration the cost of labour is high and supervision difficult; but this ignores the purpose of scattered working and the value of the benefits gained from the resulting forest structure. It is always necessary, when choosing a silvicultural system, to consider both the costs likely to be incurred and the benefits anticipated.

Because young trees stand below old ones, there are leaves and roots at every level within the stand and the soil, light, water, and nutrients are well-used. The question of whether forests of selection type have a greater biomass and are more productive than those of regular type on similar sites is difficult to answer precisely, but the studies of Assmann (1970) suggest that the differences are small (page 16).

It is frequently held that fire is less destructive in selection forests than in even-aged forests since, although the younger trees may be killed, older trees survive to regenerate the area; the management of selection forest is also said to be less disorganized by fire. It is, however, arguable that there is greater danger of fire spreading into the crowns in selection forests and that with regular crops it is possible to concentrate protective measures on to young crops where the risks are great. Which arguments are valid will depend on circumstances. The selection forests of shade-tolerant species with which central European foresters are familiar grow in a climate where there is little risk of fire, except perhaps in dense, young, even-aged stands.

It is sometimes argued as an advantage for the selection system that it is the most natural one, in that the selection type most nearly resembles natural indigenous forest; but this assumption is by no means correct. It appears that in virgin forest, both temperate and tropical, the true 'selection type' of forest, containing trees of all ages or sizes properly distributed, is exceptional and requires the careful intervention of man to maintain it.

Application in practice

One of the most important forest types in Europe is the mixed conifer and broad-leaved forest of the Vosges, Jura, Alps, and Carpathians. The main species are Norway spruce, European silver fir, and beech, with some sycamore, rowan, and occasional other species. There are several variants of the main forest type and all are typically montane, being confined to hills and the middle slopes of the mountains and rising to elevations of 1500 m.

A major feature of the climate is the long and often very severe winters with more than 4 months of snow. The progression from winter to summer and back again is sharp; allowing for a few days of spring and autumn the growing season may be as long as 200 days, but is often shorter at the higher elevations. Rain is plentiful and evenly distributed with heavy summer showers; including snow, total precipitation varies from 1000 to 2000 mm spread over 100 days in the year.

The geology consists in general of Upper Jurassic and Cretaceous rocks and the forests all occur on the same calcareous subsoil of more or less fissured, permeable rocks. In the Vosges the forest lies over a rather rich granite giving fertile soils. Occasionally the parent rocks are overlain with glacial deposits and where these are heavy, as in the Jura, there is a greater proportion of broad-leaved trees. Generally the soils are kept constantly moist by rain and snow and the flow of water from the higher slopes.

The proportion of species varies from site to site; in general that of beech is greater on sunny or south-facing slopes than on the shaded, north-facing slopes. The forest normally has a storied structure with the conifers project-ing above the broad-leaved trees of the lower storey; the range in the size of trees is striking. Gaps in the stands are filled with young regeneration. The groups of young trees are gradually thinned down to smaller numbers and even to single trees, or perhaps no tree in a group will ever succeed in reaching the upper canopy. The crowns of the conifers are very deep, forming an open upper canopy with younger trees coming on below. The shorter trees push up beside the taller, helping to restrict the crowns of the latter, and when the larger trees are felled they are gradually replaced by the shorter. Thus there are leaves at all levels and light is well-used (Ammon 1951).

The climate and soil favour the Norway spruce (which is the main species) but it roots shallowly, and where the rock is not deeply fissured or where there are not deep pockets of penetrable soil it is likely to be blown down by strong wind. The Silver fir roots more deeply, forming a massive root system which penetrates the soil and leaves it porous (Anderson 1960). Because the Silver fir grows tall it is more liable to windthrow than the beech or sycamore which form centres of stability, especially if they are located on ridges; these two species are present throughout the stands to the extent of 10 to 20 per cent by area and 5 to 10 per cent by volume. Their heavy leaf-fall becomes mixed with the needles of the conifers to give a loose litter layer.

Norway spruce does not regenerate directly beneath spruce but does appear very consistently at the base of Silver fir trees and around stumps (Ammon 1951); thus it appears that the presence of a proportion of Silver fir is essential to ensure the regeneration of Norway spruce. Conversely Silver fir regeneration comes in under spruce. Seeds of beech and sycamore readily germinate in the mixture and the seedlings only require light in order to develop. Rowan comes into the stands readily, even from long distances, being carried by birds. If natural regeneration is insufficient planting is done to maintain the desired mixture of species.

As noted, the stands do suffer wind damage; normally the oldest and tallest trees catch very strong winds and are blown down. Their loss does not completely devastate the forest but leaves it almost intact with young and middle-aged and young trees to carry on undamaged. Norway spruce suffers less from deer browsing than do the other species, especially the Silver fir. Conifer heart rot, *Heterobasidion annosum*, attacks Norway spruce but Silver fir is more resistant and the presence of the second storey of broad-leaves may reduce the incidence of heart rot.

On a favourable site the mean annual increment of a selection forest composed mainly of Norway spruce and Silver fir with some beech is 10 to

11 m^3/ha per year, and the proportion of trees exceeding 50 cm in diameter at breast height can be very high. Thus the irregular spruce, fir and beech forest is a stable, productive, and permanent selection forest type. At the higher elevations and in areas reserved primarily as protection forest, it is commonly treated by the single tree selection system. At lower elevations and in more sheltered situations, as in the Vosges and Jura of France, it is worked by the uniform system with floating periodic blocks (see page 111).

The selection system is being applied in the United States of America, particularly in the northern hardwood association (Tubbs *et al.* 1983). The selection type of forest is often maintained to provide amenity and facilities for recreation in the immediate neighbourhood of towns and mansions. In protection forests the system may be used where seeding is rare or spasmodic and it is difficult, if not impossible, to systematize regeneration to any extent. Similarly, it is also useful on broken, rocky or precipitous ground where continuous regeneration is out of the question and advantage has to be taken of scattered regeneration appearing in places where there is sufficient soil to support tree growth.

The modern selection system emerged during the period between 1880 and 1920 and the sequence of events was somewhat as follows. During the eighteenth century the forests of Switzerland became much depleted by wasteful methods of working. The previous unregulated selection fellings had caused deterioration of the forest. By the beginning of the nineteenth century action was needed to remedy matters. At this time great progress was being made in Germany with clear cutting followed by the formation of compact even-aged stands by planting. From about 1920 onwards for a time, almost all Swiss foresters had their training in Germany, so were influenced by the new German ideas and practices which were, in consequence, introduced into Switzerland. During most of the nineteenth century forestry in Switzerland remained under the influence of foreign teachers and models. Swiss foresters gained much from Germany but their uncritical acceptance of theories and practices developed under different conditions of climate, terrain, and land tenure led to problems (Knüchel 1953). Hence when Karl Gayer (1880) declaimed against the wholesale growing of Norway spruce and advocated the adoption of more natural methods he found willing listeners in Switzerland. Arnold Engler (born 1858, died 1923), Director of the Federal Forest Research Institute, relied particularly on the writings of Gayer in building on an already existing trend toward uneven-aged forestry. Then Henri Biolley (born 1858, died 1939) developed the methods that the French forester Adolphe Gurnaud (born 1825, died 1898) had devised for estimating increment and controlling fellings, and put them into practice during almost 50 years of work in the forest of Couvet and other forests of the Val de Travers.

One of the men who established a practical basis for the selection system was Walter Ammon (born 1878, died 1956). He studied at the Technical High School in the time of Arnold Engler and graduated in 1902. In 1912 he moved to the Thun district which includes the selection forests of the Emmental. 'Das Plenterprinzip in der Waldwirtschaft' appeared in 1937 and ran to three editions. In it Ammon described the ideal structure of the managed selection forest and the details of its silvicultural treatment.

The continued strength of interest in the 'natural' approach to silviculture (page 11) makes it certain that where conditions of climate, terrain, and species are similar to those in parts of Switzerland the selection system or some modification of it will eventually be adopted.

14

The group selection system

Fr. Jardinage par bouquets; Ger. Horstweiser plenterbetrieb; Span. Metodo de entresaca por bosquetes

General description

The typical form of selection fellings in which single trees are removed is more suited to shade-bearing than to light-demanding species, since the gaps created by the removal of single trees are too small for the regeneration of light-demanders or for the smaller trees to escape suppression from the larger ones standing over them. Hence if the selection system is to be applied to light-demanding species it is necessary to fell trees in groups in order to create gaps of sufficient size to enable regeneration to establish itself. Such fellings are known as group selection fellings and the modification of the selection system under which this type of felling is employed is termed the group selection system.

Application in practice

Provided that Dwarf mistletoe (*Arceuthobium* sp.) and Conifer heart rot (*Heterobasidion annosum*) are not prevalent, the group selection system has been applied in the United States of America to the north-western Ponderosa pine forest type where amenity is important (Barrett *et al.* 1983), the Ponderosa pine and Rocky mountain Douglas fir types (Ryker and Losensky 1983), and to the south-western Ponderosa pine forest type (Ronco and Ready 1983). In the Sierra Nevada, Ponderosa pine grows together with California white fir, Sugar pine, coastal Douglas fir, Incense cedar, and California black oak (Laacke and Fiske 1983) and the group selection system has also proved useful there. If Dwarf mistletoe and Conifer heart rot are prevalent the system must be ruled out because these damaging agents are perpetuated by it.

Conversion of regular to irregular high forest

By conversion is meant a change from one silvicultural system to another (page 225). Several methods have been devised to convert forests composed of regular stands worked under the clear cutting system to an irregular structure, and the group selection system has been the preferred one. The purpose of this section is to describe some examples of conversion to show how it may be achieved.

In the Ardennes region of Belgium small areas of forest composed of regular stands of Norway spruce and other conifers, including Douglas fir, worked under the clear cutting system are being converted to the group selection system using a method devised by Turner (1959). The object is to diversify the uniform stands by introducing beech, Silver firs, and other species to develop an irregular structure (Anderson 1949; Penistan 1960). Norway spruce is retained on sites where it is growing and regenerating well but the proportion of it in the new crop is being reduced.

Turner introduced and developed the method whilst he was the forest officer in charge of the state forest of Grand Bois near Vielsalm in 1933. The spruce forests of the High Ardennes commonly lie at elevations of 450 to 550 m. They are found on brown forest soils of moderate depth and quite free drainage. The terrain is level or slightly undulating. The original forest of the region had consisted largely of beech and, because it had been over-exploited, Norway spruce and other conifers were introduced from about 1850 to 1860.

At Vielsalm the process of conversion normally begins at 60 years when the trees have reached the end of their maximum growth in height. Because the stands have been regularly crown-thinned since the age of 25 years the dominants have well-developed crowns. A preparatory felling is made to isolate the best trees. At this time regeneration of Norway spruce, some rowan, and birch appears commonly in the stands.

Six years later, in each hectare of the stands to be converted, two circular plots with diameter of 36 m and area of 10 ares (or, more recently, one plot of 20 ares) are aligned in a direction south south-west to north north-east. In each plot the middle zone of trees over a width of 10 m and in a direction at right angles to the prevailing south-westerly wind, is heavily thinned to let in light from the south. Four or five trees with stems of high quality are left standing in this zone. In the north the suppressed trees are retained; in the south they are removed (Fig. 29).

In the following year beech or Silver fir is planted in closely spaced lines. Natural regeneration of Norway spruce (and Douglas fir if it is present in

Figure 29.   Conversion of regular to irregular high forest. A clearing containing young beech made in a pure, regular stand of Norway spruce. Seen from the south-west. Forêt de Grand Bois, Vielsalm, Belgium.

the existing stand) develops in the middle zone of the plot. When this and the planted beech and Silver fir are established, the groups are extended northwards by thinning repeatedly to isolate the dominant trees. To the east of the regeneration suppressed trees are removed to let in more light, and Douglas fir may be planted as the canopy becomes more open. Elongated groups of regeneration are established through the stand and eventually they coalesce.

At Vielsalm the aim now is to produce a crop composed mainly of Norway spruce (up to 80 per cent), Beech and Silver fir (each about 10 per cent), and Douglas fir. In the communal forest of Tellin near Rochefort, Japanese larch is also introduced when the conversion is well-advanced, into clearings made by fellings in the old crop (Roisin 1959).

The rotation age is normally 90 to 100 years and the aim is to complete the conversion in about 30 years. To organize and control the work the forest is divided into three blocks each comprising four compartments; the felling cycle is 12 years. There is a good network of roads in the forest of Grand Bois. Extraction is done by horses and little damage is done to the young trees. Although numbers of Red and Roe deer (*Cervus elaphus* and

Capreolus capreolus) in the Ardennes have been reported as being high in the past, fencing is evidently not needed to protect the regeneration now.

Another example of conversion from clear cutting to the group selection system comes fom the Tavistock estate in south-west England. Here the method entails systematic division of regular stands of conifers into small, self-contained units of similar size so that all parts of each stand are being converted simultaneously. These units of regeneration are marked on the ground at fixed intervals through existing stands, and the area of each is determined by the growing space required to grow trees of the succeeding crop to a specified minimum size (Hutt and Watkins 1971; Bradford 1981).

The warm, moist climate and generally fertile soils of the south-west peninsula of England favours rapid growth of Douglas fir, Western red cedar, Western hemlock, Coast redwood, and Norway spruce and also of broad-leaved species, including *Nothofagus procera*. Although the patterns of height growth differ between species, measurement of the diameter increment of conifers in the area has shown that trees of 60 cm diameter can be grown in 50 to 60 years depending on the capability of the various sites. The crown projection area for dominant trees with this stem diameter is 36 m^2.

The conversion from a regular to an irregular structure is being done in nine stages with a felling cycle of 6 years, so each self-contained unit of regeneration consists of nine square plots, each 6 × 6 m, giving a total size of 18 × 18 m or three ares. There are 30 of these units per hectare. The rows of units are aligned at right angles to a main extraction route and timber from thinnings and fellings is extracted by tractor along tracks lying between the rows.

The process of conversion begins in regular crops at or near the stage of first thinning. All the trees in the central plots of the regeneration units are felled, the slash is removed, and any coppice shoots or saplings of unwanted species are cut. Each cleared plot of 36 m^2 is then restocked with nine transplants of one species. The other eight plots are thinned to free the crowns of selected trees, provide light to the young trees, and stimulate diameter growth of trees in the overstorey. It is also necessary to ensure that tractors hauling logs have an unimpeded route to the main extraction road.

At year 6, the second plots in the regeneration units are felled and restocked with the same or a different species if mixtures are desired. The remaining seven plots are thinned to the desired basal area, or volume. This sequence continues until year 54 when each unit will contain:

(1) four selected trees aged 36 to 54 years which form the upper canopy. These will have been high pruned;

(2)   three groups of trees aged 18 to 30 years which are being thinned (on the best sites there will be a fourth group ready for thinning at 12 years;
(3)   one group of trees 6 years old.

During thinnings trees with straight stems, small branches, and rapid diameter growth are selected for high pruning, which is done in stages to a height of 7.5 m so that the basal log, which is normally 6 m long, has a wide zone of knot-free timber.

It is essential for the survival and good growth of the new crop that rabbits (*Oryctolagus cuniculus*) and deer are controlled. If some broad-leaved coppice shoots and saplings are retained in the lower storey of the existing crop, Roe deer tend to browse and fray these rather than the conifers.

The conversion from regular to irregular forest began at Tavistock in 1961 and the process is about half-way towards completion. When the first and second plots in each unit of regeneration were felled they proved suitable only for the shade-enduring Western red cedar and Western hemlock; but when the third plots were felled and three thinnings had been made in the other plots, more side and overhead light reached the young trees and a greater range of species could be planted, especially Douglas fir, *Nothofagus procera*, and *N. dombeyi*.

The advantages of this method of conversion appear to be:

1.   Fewer trees are planted per hectare than on clear felled areas. Their survival is high because the overhead shelter protects them from damage by late spring frosts and the season of planting can be extended so avoiding losses from spring droughts. Fewer and less intensive weedings are required than on clear cut coupes. All told the cost of establishing the crop is normal for the region.
2.   There is a flow of revenue throughout the period of conversion and the existing growing stock is being improved in tree size, quality, and value.
3.   The more open canopy and varied heights of the trees adds to the aesthetic value of the stands being converted. The bird population has become more diverse and numerous.

The disadvantages are:

1.   The silviculture, yield control, and marketing of produce are demanding. The future value of the irregular crop will depend on the proportion of Douglas fir that can be established, and it cannot be introduced until the third felling cycle.
2.   The forest workers must be trained in the intensive methods of working, but the demand for labour is stable so that a trained work-force can be built up.

3. The method has not proved suitable for converting regular stands of beech or Sitka spruce, and it is difficult to apply in regular stands of Douglas fir and Western hemlock because of persistent branching. Under the favourable growing conditions of south-west England it is often necessary to thin every 3 years.

The examples from Belgium and England described above impose a formal procedure on the conversion from a regular to an irregular condition. At the British Forestry Commission's Glentress forest in south-central Scotland a trial of conversion from clear cutting to the group selection system began by attempting to establish the new crop in small clearings arranged at specified distances apart; but this was subsequently relaxed in favour of a more flexible procedure using larger clearings.

The trial conversion at Glentress began in 1952 on an upland site moderately exposed to wind (Blyth 1986). The objects are to establish an irregular forest of mixed conifers and broad-leaves which is resistant to wind and snow, gives protection to the soil, and produces timber of high quality and value. The site of the trial extends from 240 to 560 m with most of it above 380 m. The soils are mainly of brown forest type, and the annual precipitation is around 1100 mm. It is probable that the original forest was broad-leaved but the present crops comprise Douglas fir, Japanese and European larch, Norway spruce, and Scots pine varying in age from 60 to 85 years. A maturing stand of Sitka spruce adjoins part of the trial area.

The total of 117 ha has been divided into six blocks with three compartments in each, the felling cycle being 6 years. Spruce, both Norway and Sitka, are the main species in the new crop, with larch, and Western hemlock on the higher ground, and Grand fir and Douglas fir on the lower sites. Beech, sycamore, and other broad-leaves have been planted throughout. It is intended to complete the conversion to an irregular structure in 60 years.

At first the clearings ranged in size from 2 ares at the higher elevations to 20 ares on the lowest; these proved too small and the range now used is from 0.1 to 0.25 ha, that is, with sides or radii about twice the height of the surrounding crops—which are from 18 to 30 m. In 1987, with five of the ten felling cycles completed, natural regeneration of sycamore is plentiful and in some parts natural regeneration of Sitka spruce is sufficient to form acceptable crops. Rabbits, hares (*Lepus* sp.), and Roe deer have damaged beech and Silver fir; Western hemlock was damaged by deer and snow.

In the same part of Scotland, there are two more examples of conversion from regular to irregular forest. Conversion of stands surrounding the mansions at Bowhill near Selkirk and Eildon near Melrose, began in 1926.

Since 1956 there has been continual establishment of groups of conifers and broad-leaves in various sizes of clearings, usually on a 6-year cycle. Groups of the upperstorey trees have been retained to accentuate the group selection structure which now extends over more than 300 ha and is the largest in Britain.

Conversion of unmanaged forest to the group selection system

In the lowlands of Britain, particularly in England and Wales, there are areas of unmanaged or degraded forests, usually resulting from fellings made during the first and second world wars and intervening periods of neglect. Since the 1950s methods have been devised to restore them to productive forests (see page 225), and one of these methods is conversion to the group selection system.

General description

The purpose of the method is to convert small areas of unmanaged broad-leaved forest to productive stands so as to maintain permanent cover of an irregular type, which consists of more or less even-aged groups of trees (Garfitt 1984). The existing stands consist of a mixture of unthinned trees varying in age and quality from over-mature standards with large crowns, to dense thickets of natural regeneration. The most common species are oak, beech, ash, birch, and sycamore with some Scots pine, European larch, and Douglas fir. The amount of the existing crop that is retained is related to the proportion of broad-leaves that it is intended to retain as a permanent constituent of the forest. This will vary in individual cases from about one-third where the financial returns are important, to the whole area of the forest where only amenity considerations are taken into account.

The first step is to examine the existing growing stock to locate groups which by their quality can be expected to improve if carefully thinned. The size of these groups will usually be small—from 0.20 to 0.30 ha but larger groups do occur. All except the selected groups are clear felled and the ground is restocked with groups of approximately equal size comparable to that of the selected groups. A suitable diameter for the planted clearings is 18 to 20 m which will give space for up to 70 plants at 2 m square spacing. Larger cleared areas of up to 0.6 ha are planted with several groups, each usually consisting of one species.

All the stands are now tended and thinned. The importance of cleaning thickets of natural regeneration is emphasized by Garfitt (1979, 1980) whose procedure for releasing potential crop trees and removing badly-shaped ones is based on the growing space needed by trees of the final crop as they approach exploitable size. Thus if the mature trees will have an average crown width of 9–10 m, two vigorous young trees with straight stems and small branches must be released every 9 or 10 m when the crop is 2–3 m tall (see page 177). Jobling and Pearce (1977) give dimensions for the space required at different ages for the rapid diameter growth of oak. Brashing is necessary on private properties where field sports, especially pheasant shooting, is a consideration.

When the forest has been satisfactorily stocked, conversion to the group selection system begins. The area to be felled and regenerated in 1 year is given by the time taken by the various species to reach exploitable size and the number of years elapsing between successive visits to the same groups —called here the treatment cycle to distinguish it from the strict felling cycle (see page 166). To avoid small fellings and plantings the treatment cycle should lie between 5 and 10 years. Thus, if a mainly broad-leaved forest consisting of ash, sycamore, Wild cherry, and Small-leaved lime is to be worked on a rotation of 60 years and the treatment cycle is 10 years, one-sixth of the total area will be felled and replanted every 10 years, and this area will consist of a number of groups. During the process of conversion to an irregular structure the first felling and replanting is done at the tenth year after conversion begins, the fastest growing crops being chosen. By the seventieth year a full series of age classes from 1–10 to 51–60 years will have been established. On completion of the conversion there will often be some groups of over-aged trees; this can be turned to advantage if particularly fine trees are reserved earlier to put on increment and value.

Felling is controlled by area, the units being the groups. This has the advantage of being clearings evident on the ground so avoiding overcutting. If desired, the increment of the growing stock can also be assessed by recurrent inventory of permanent sample plots. The amount of damage done to young trees during felling and extraction can be controlled because the fellings are equivalent to limited clear cuts. Extraction is confined to tractor paths leading straight to the main extraction routes.

Advantages and disadvantages of the group selection system

The advantages are:

1. It is very flexible and particularly well-suited to small forests where intensive working and close supervision are possible.

2. It can accommodate a wide variety of species ranging from light-demanders, through the moderate shade-bearers to the shade-enduring species. Site variations can be matched by species; some advantages may also be gained from placing certain species in adjacent groups.
3. Application of the principle of improving the growing stock by selecting and favouring the best trees will increase the quality and value of the timber produced. New species, better provenances, and improved cultivars can readily be introduced.
4. The range of produce is wide and can include some Christmas trees or species with decorative foliage.
5. The varied habitats on the ground and at different heights in the canopy favour plants, birds, and mammals. With appropriate layout and treatment of the groups of trees, game birds and sporting interests can be accommodated.
6. The mosaic of open glades, thickets, and tall trees is attractive to walkers and riders. The group selection system does not vary much in its external appearance so it forms a permanent background of amenity to towns, historic buildings, and areas of scenic beauty popular with the public.

The disadvantages are:

1. Silvicultural skill is needed in selecting groups of trees in the existing crop for retention and, where several new species are to be introduced, in deciding on suitable sizes and locations of the clearings for each.
2. The work of felling, extraction, planting, and tending is scattered and on a small-scale. It must be done under experienced and knowledgeable supervision. The costs are likely to be higher than those in a comparable area of regular forest (but see page 169).
3. Control of vermin is more difficult than in regular forest and damage done can be more severe.

Concerning the likelihood of damage by wind, Garfitt (1984) suggests that it is reduced in group selection forest because the groups are constantly exposed to wind and will therefore remain stable.

15

Accessory systems

The formation of two-storied high forest and high forest with reserves provide examples of accessory systems (Troup 1952, p. 133) because they may arise from various other systems and are not dependent on any particular form of regeneration.

Two-storied high forest

Fr. Futaie à double étage; Ger. Zweihiebiger hochwald; Span. Monte alto de dos pisos

General description

Two-storied high forest is composed of an upper and a lower storey of trees of seedling origin, growing in intimate mixture on the same site. Generally two species are involved, the upperstorey usually comprising a light-demander under which a shade-tolerant species can grow without being suppressed when introduced at a later date.

Trees of the upperstorey may arise through natural regeneration or planting. They are treated as an even-aged crop in terms of stocking and thinning until they approach middle age, when a heavy thinning takes place. At this stage the crop is underplanted with the second, shade-bearing species which forms the lower storey. Sometimes the lower storey is established through natural regeneration before or soon after the heavy thinning is done, or in other cases direct seeding into the thinned stand is used. Both stories are allowed to grow up and subsequent thinnings are done in the lower storey. Both stories may be felled together, or the upperstorey is removed in one or a series of fellings to leave the lower as an even-aged maturing crop. Two-storied high forest is frequently limited to only one rotation because of the difficulty of forming a second understorey under the previous shade-bearing one.

In Europe the most common reason for using two-storied high forest is to encourage diameter increment on selected upperstorey trees whilst protecting the soil with the lower storey. As long ago as 1830 the Hanoverian forester von Seebach obtained light increment (Ger. *Lichtwuchsbetrieb*) on beech by removing 50 to 60 per cent of the growing stock from stands 70 to 80 years old in response to a demand for large timber which the growing stock was ill-equipped to provide (Köstler 1956, p. 325). Nowadays in Europe this objective is most frequently met with Scots pine underplanted or under-sown with Silver fir, or Scots pine and oak underplanted with beech.

Hiley (1959) experimented with the two-storied high forest system on the Dartington estate in south-west England. He wished to grow Japanese larch to large diameters but noted that in well-stocked stands current annual diameter increment culminated quite soon. In order to maintain rapid growth in diameter the trees need large crowns and this requires virtual isolation at an early age. Hiley decided that underplanting with a shade-bearing species would make fuller use of the site whilst allowing rapid growth of the Japanese larch. He also hoped to exploit a strong local market for mining timber with the produce from the heavy thinning of the upper-storey, avoid unprofitable later thinnings, and combine the advantages of an irregular type of forest with the economies of the clear cutting system. Site protection and weed suppression were to be achieved by means of the lower storey of shade-bearing species which would be tended as an even-aged crop.

In 1955 a compartment of Japanese larch was thinned to 120 stems/ha and underplanted with Western hemlock, Western red cedar, and Douglas fir. The larch was 25 years old and had received four previous thinnings (Hiley 1959), the last having the specific object of improving root develop-ment and stability to wind. Howell *et al.* (1983) summarized the results of this trial. The larch was clear felled at 48 years, by which time the crowns had all but closed, the mean diameter at breast height had reached 47 cm and one-third of the timber was suitable for boat building. Where Douglas fir formed the lower storey it had been suppressed but Western hemlock and Western red cedar had thrived. This was a fertile site; in other com-partments on poorer soils, underplanted at the same time or later, the Japanese larch had more typical stem diameters of 33–36 cm. These compartments had been less carefully prepared by thinnings and the under-planting had been delayed until the upperstorey was 30 years old. Although Hiley achieved his silvicultural and marketing objects in the best two-storied stand, his financial predictions proved optimistic.

Hiley (1969) saw a further benefit from two-storied working, namely, that it enables the manager to adjust the distribution of age classes within a forest. Where there is an excess of young age classes, early maturation of the upperstorey provides a final crop sooner than is possible in a uniform stand.

Moreover, the lower storey helps to fill the gap in the distribution of size classes which may otherwise be caused by not having any land to plant.

Historically it was thought that some species were 'incapable of maintaining soil fertility' and that growing a lower storey of a different, usually broad-leaved, species would help to protect the soil from degrade (see page 13). Site improvement has been reported from the Sauen district of Germany where Scots pine was underplanted with beech and Locust tree (Flöhr 1969). On the better soils the volume increment of underplanted pine was reported to be 79 per cent higher over a period of 13 years than in pure pine. However, as with Hiley's experience at Dartington, the poorer sites gave little gain and did not justify establishing a lower storey.

It has been suggested that the total production can be increased by growing two different crops simultaneously. They place different demands on the soil, and the canopy structure allows development of a greater leaf area for photosynthesis. Furthermore there is no period when the site is not being utilized, as occurs with the clear cutting system. Increases in yield can be obtained but where soil moisture is limiting or site fertility only moderate both crops may suffer and total production declines. Zundel (1960) found that Scots pine underplanted with Silver fir gave a volume midway between that of pure Scots pine and pure Silver fir.

Advantages and disadvantages

The system offers several advantages:

1. The lower storey protects the soil and gives freedom to thin the upperstorey heavily, so encouraging the best stems to put on light increment. This rapid growth can shorten the rotation whilst producing 'mature wood' of good quality, thus providing two financial benefits.
2. An upperstorey provides protective cover for establishing shade-bearing species which are subject to frost damage.
3. Where a mixed stand is required but the growth rates of the species differ, the system can be used to give the slower-growing species an early start.
4. There may be early returns from the heavy thinning at a time when the trees are still at a moderate size.
5. It can be used to effect a gradual change of species and produce through the lower storey.
6. In its early stages the appearance of two-storied high forest is pleasing.

The disadvantages are:

1. In windy climates the sudden exposure of the upperstorey by the heavy thinning can result in windthrow. The trees to be retained must be

prepared for this several years in advance so that they develop strongly tapered stems and good root systems. This implies early commitment to the establishment of a lower storey rather than reliance on the system to resolve a silvicultural or management difficulty at some later stage.

2. When the upperstorey trees reach the age of felling, damage can be done to the lower storey unless care is taken. Appropriate thinning of the lower storey will allow felling into prepared gaps (Howell *et al.* 1983).

3. Two-storied high forest is more difficult to manage than a regular one, both in terms of balancing the growth of both stories and in carrying out regeneration and tending operations. It is, however, considerably simpler to manage than the selection system whilst sharing some of its advantages of full site use, soil protection, weed suppression, and visual appeal.

Application in practice

Two-storied high forest is used in the irrigated stands of *Dalbergia sissoo* and *Morus nigra* at Changa Manga in the Punjab region of Pakistan (Champion *et al.* 1973; Khattak 1976). *D. sissoo* is a pronounced light-demander and *M. nigra* is shade-enduring. The lower storey of mixed *Dalbergia* and *Morus* comprises 40 per cent of the growing stock and is cut on a rotation of 15 to 22 years. The upperstorey of pure *Dalbergia* is retained for three rotations of the lower storey; it comprises 60 per cent of the growing stock.

Until 1927 *Dalbergia* was regenerated by direct seeding; since then stumped plants, in which both root and shoot are pruned have provided a surer means of restocking. As already noted on page 83, seedlings of *Morus* sprang up from seed dispersed by Red-breasted starlings (*Sturnus roseus*) but stumped plants are now also used to ensure full stocking. Originally all the irrigated stands were formed to provide fuelwood but with the use of two-storied high forest it became possible to produce sawlogs on the better sites.

Miegroet (1962) studied small forestry properties in Belgium and calculated minimum areas for regeneration, tending, and economic management of stands. He considered that for areas of 5 ha or less, two-storied high forest was a promising system, using light-demanding species with rapid growth and short rotations, such as ash, sycamore, Wild cherry and even tree willows as the upperstorey and Douglas fir, Western hemlock, and Grand fir in the lower storey. James (1982) listed oak, ash, and larch as suitable upperstorey species in Britain, with beech, Western hemlock, and Grand fir in the lower storey.

The application of the system may be limited by the need for relatively fertile sites with good supplies of water. Without these the total yields will probably be less than that of the most productive species in the mixture when grown as a pure even-aged crop. Recently it has appeared possible, but by no means certain, that two-storied high forest may have a use in regions troubled by pollution in providing a protective intercepting upper-storey for susceptible species, such as Silver fir.

High forest with reserves

General description

High forest with reserves (Fr. Réserves; Ger. Überhalter; Span. Arboles reservados) is produced by leaving selected stems of an old crop standing over a young crop established by regeneration from the old one. These stems which are known as 'reserves' or 'standards' may be retained, scattered singly or in small groups, for the whole or part of a second rotation. In two-storied high forest a second crop is introduced under an existing one, and this is the feature distinguishing the two systems.

The chief object of this system is to retain selected stems to put on light increment and produce large-sized timber which is used in boat building, constructing, and restoring buildings of special significance, and to produce veneers and panels of special quality to decorate the interior of public buildings and private houses. Thus reserves are normally of species in demand for such purposes, such as Scots pine, European larch, oak, and beech.

Reserves may also act as seed bearers to fill blanks in young crops, or they may be an insurance against destruction of the young crop by fire. Another important reason for retaining reserves is to create amenity by producing stands with a natural appearance. In this case the reserves are often situated alongside rides or roads so that they can be seen and enjoyed by walkers or motorists.

The selection of reserves requires care. Large trees suddenly isolated are liable to be thrown by wind or in certain cases become sickly and die. Reserves should be of a windfirm species and have well-tapered stems, in addition to strong crowns and root systems. In some cases it is customary to prepare trees for isolation by selecting prospective reserves many years ahead so that they may receive special attention in thinnings. To prevent suppression of the young crop, reserves should have long stems which carry the crowns well above the ground and the crowns should not be too heavy or spreading.

Oak is often retained as a standard, but is liable to become stag-headed and produce epicormic branches when isolated. In the Spessart highlands of West Germany it is customary to retain 25 reserves of oak per hectare over oak and beech regenerated by the uniform system (page 117). Whenever possible these reserves are surrounded by beeches to shade the stems and prevent formation of epicormic branches. These old oaks produce veneer logs of high quality.

Beech is generally an unsuitable reserve, as it has a heavy crown and is also liable to sun-scorch when suddenly isolated. In Switzerland, however, beech reserves with well-elevated crowns are retained over mixed crops regenerated under the irregular shelterwood system, the height of the young crop protecting the beech stems from sun-scorch. In the Søro Academy forest in Denmark selected beech seed bearers become reserves standing over young crops regenerated under the uniform system. These large trees are felled in winter into crops in the thicket stage with remarkably little damage done to the young trees.

Scots pine reserves with tall stems and full crowns are also reserved for a second rotation to produce timber of large dimensions and high quality. The number of reserves varies from 20 to 40/ha and they are retained for between 160 and 200 years (Köstler 1956, p. 323). Because the crowns are usually carried high off the ground there is less interference with the young crop than might be expected.

European larch also makes a good reserve or standard on good sites. Generally, in the mixed conifer forests of central Europe, it is often the custom to leave reserves of European larch and Scots pine over naturally regenerated mixed crops for part or the whole of a second rotation. As an example, reserves of Scots pine are retained over mixed crops of Silver fir, Norway spruce, and Scots pine regenerated under the wedge system at Villingen-Schwenningen in West Germany (see page 149).

European silver fir and Norway spruce are generally unsuitable as reserves but they are kept for amenity, especially near to rides and roads.

Advantages and disadvantages

These are similar to those of two-storied high forest. Concerning the danger of windthrow, instances have been recorded in which up to 90 per cent of Scots pine reserves have been lost (Troup 1952, p. 125). This damages the main crop also. Suppression of the main crop is a potential danger, especially when the crowns of the reserves are low and spreading. If the reserves are not expected to remain for the whole rotation they may be concentrated near extraction routes so that they can readily be felled.

Application in practice

There appears to have been a decline in the use of high forest with reserves in Europe during the past 20 years. In France it is not much used whereas in East and West Germany it is more plentiful. In Britain, if a proportion of large or large-sized timber is needed a compartment should be set aside and all the trees in it allowed to remain as a self-contained crop for all or part of a second rotation, if necessary with the aid of an understorey to protect the soil. Suitable thinnings would be done to ensure light increment. A good example of this can be seen in Reelig Glen near Inverness in northern Scotland. Douglas fir is a major constituent of a mixed and somewhat irregular stand growing on a fertile, moist, and sheltered site. This stand of reserves is also popular with the public because of the varied habitat and range of bird life.

16

The coppice system

Fr. Taillis, taillis simple; Ger. Niederwald; Span. Tratamiento de monte bajo

General description

This chapter deals with the system known as 'simple coppice'—a term sometimes used to distinguish it from coppice with standards—together with certain modifications of it, including pollarding.

Production of shoots

The coppice system involves reproduction by stool shoots or suckers. When felled near ground level most broad-leaved species, up to a certain age, reproduce from shoots sent up from the stump or stool. These shoots, known as stool shoots or coppice shoots, arise either from dormant buds situated on the side of the stool at or near ground level or from adventitious buds arising from the cambial layer round the periphery of the cut surface. The former are more firm on the stool, and are therefore more important for purposes of regeneration. As a rule many shoots arise from each stool but each year some of the smaller shoots die until from 2 to 15 remain, with the result that coppice has a characteristic clumped appearance.

As a general rule coppice shoots are not produced by conifers, but certain species, notably the Chinese fir (Lin 1956), the Coast redwood of California (Barrette 1966), *Tetraclinis articulata* in North Africa (Stewart 1980), *Pinus canariensis* and *P. oocarpa* (Matte 1965) are exceptions to this rule. Among broad-leaved species the age up to which stool shoots are produced, the number and vigour of the shoots, and the vitality of the stools vary considerably.

In most species, stools of large sizes do not coppice strongly and hence for the production of stool shoots it is usually necessary to fell trees at an age of not more than 40 years, and in some cases considerably less; Sweet chestnut and lime on the other hand coppice to some extent up to an age of

100 years. Among European trees that coppice freely are ash, oak, hornbeam, sycamore, lime, alder, hazel, and Sweet chestnut. Birch and beech coppice less vigorously. Aspen, although it produces suckers freely, does not coppice. Oak and hornbeam stools retain their vitality for a long time, lasting through many successive cuttings; those of ash, birch, sycamore, and beech are less enduring. Several species and cultivars of willow are worked as coppice on short rotations and respond well to such treatment.

North American trees that coppice freely include Red alder, American plane, Western balsam poplar, Sweet gum, and Yellow poplar. Many tropical and sub-tropical trees coppice vigorously; among the *Myrtaceae*, for instance, *Eugenia jambos*, and some species of *Eucalyptus* show phenomenal coppicing power. Some provenances of *E. camaldulensis* coppice well for six or more successive cuttings. *E. microtheca* and *E. gomphocephala* also coppice well in arid and semi-arid areas. *Azadirachta indica* is a member of the *Meliaceae* which coppices strongly in arid areas. The tamarisks and several leguminous trees and shrubs coppice with vigour. Examples among the *Mimosoideae* are *Acacia senegal, Prosopis juliflora*, and in the humid tropics *Calliandra calothyrsus* and *Leucaena leucocephala*. *Cassia siamea* is a species of the *Caesalpinioideae* which continues to yield well after four or five successive coppice cuttings. *Tectona grandis* usually coppices less vigorously than these species.

Some species regenerate freely from suckers. Examples among trees growing in Europe are White and Grey poplar, aspen, Wild cherry, English elm, and Grey alder; in the Mediterranean region Holm oak; and in North America Big tooth aspen, Quaking aspen, and Locust tree. The production of suckers is stimulated not only by felling the parent tree but also by exposing and wounding the surface roots or grubbing up the stump (Banti 1948). Many species that regenerate from suckers produce stool shoots at the same time, notably *Acacia saligna, Ailanthus altissima, Ziziphus mauritania*, and Locust tree.

Method of cutting

Coppice is usually cut close to the ground, the stools being given a sloping surface which is trimmed smooth with chain saw, bow saw, or axe to prevent water from settling. The object of low cutting is to cause the shoots to appear at ground level and thus enable them to form independent roots. In the Mediterranean region in hot dry localities the Holm oak is sometimes cut below ground level, the stools being covered with earth. On land liable to inundation coppice is generally cut several centimetres above ground level; this is common practice with willows on riverain land and with alder in swampy situations. In some of the *Shorea robusta* forests of northern India, particularly in dry localities or dry seasons cutting close to the ground could result in complete failure, whereas cutting several centimetres above ground level resulted in successful regrowth of coppice shoots. The reason is that the stool dries for a few centimetres below the cut surface and by cutting near ground level the dormant buds on the stool are all killed, whereas if cutting is done several centimetres higher the stool does not dry sufficiently far down to affect the buds around the base of the stool (Troup 1952, p. 130).

Season of cutting

Coppice is usually cut during the dormant season. In temperate regions, early spring before the buds begin to swell is considered the best time. If cutting is done early in winter there is risk of the stools being damaged by winter cold through separation of the cortex from the wood. In practice the season of cutting is also governed by the availability of labour. In oak tan-bark coppice, cutting was done during May and early June, after the growing season had commenced; at this time the bark could easily be stripped and it also contained a higher pecentage of tannin than during the dormant season. If cutting was delayed too long, however, it could result in late frost damage during the following autumn, owing to continued growth of shoots and delayed onset of dormancy. Alder coppice in swampy situations is best cut during hard weather in winter, when the ground is firm. In Tanzania experiments done on *Cassia siamea* showed that the most favourable time for cutting was in May (Anon 1948) whereas in the Sahel region of Africa with a mean annual rainfall of 300 mm, *Diospyros mespiliformis* coppiced readily if cut just before the onset of the rains (Anon 1950).

Rotation and produce

Coppice is essentially a system for the production of fuelwood and small or medium-sized material up to pole size, but not for the production of large timber. The rotation and number of stools varies from 1 year with 40 000 stools/ha in the case of basket willows, to 30 or 40 years with 200 or 500 stools/ha for the production of posts, poles, and firewood billets of large size. Apart from firewood of all sizes from small sticks for domestic purposes to large billets, coppice furnishes according to the rotation adopted, material for basket work, pea and bean sticks, hoops, hurdles, fascines, fencing, vine and hop poles, handles for tools and implements, mining timber, pulpwood, posts and poles for many purposes such as telephone and electric power transmission, scaffolding, and a large variety of other produce. Among examples in Britain may be mentioned Sweet chestnut (rotation 12 to 16 years with 800 to 1000 stools/ha) for hop poles, cleft pale fencing, and fence posts, pulpwood and firewood (Crowther and Evans 1984; Rollinson and Evans 1987). *Eucalyptus microtheca* is grown in the irrigated coppices of the Gezira in Sudan to produce fuelwood and poles; the rotation is 8 to 10 years and the number of stools 1600 to 1800/ha. In Brazil, species of *Eucalyptus* are coppiced on a rotation of 7 to 10 years for the pulp and paper industry, for the production of charcoal for the iron and steel industry, and for the fibre board and particle board industries (Jacobs 1980). Ghosh (1983 in Spears 1983) has shown that a small stand 20 m square of *Leucaena leucocephela* planted at 1.0×0.5 m and coppiced biennially can supply a family's needs for fuelwood and fodder. At one time oak coppice was extensively grown in Europe for the production of tan bark, the rotation being for the most part 15 to 20 years. This has now ceased and the conversion of oak coppice has been in progress for some time (see page 236).

Working coppice

Coppice is usually worked by the method of annual coupes by area. The rotation is determined primarily on the basis of the size of material required. The area is then divided into annual coupes in numbers equal to the number of years in the rotation, and one coupe is coppiced each year. Where the maximum annual production on a given forest area is of special importance, as in the case of firewood, the rotation of maximum mean annual increment of volume or fresh weight should be determined. Rollinson and Evans (1987) described the procedure for constructing a simple growth model or yield table for Sweet chestnut coppice in southern England. This gives volumes and fresh weights for a range of ages and mean diameters. By adopting the rotation of maximum mean annual volume or weight of wood over the whole forest the total area required to produce a given out-turn of firewood each year may be reduced to a minimum.

The arrangement of coppice coupes should be such that produce may be removed from a recently felled coupe without passing through other coupes and doing damage. This requires a good system of roads and rides. To economize in roads, long narrow coupes are often formed, with roads running along the ends of them; on hill slopes similar long narrow coupes run vertically up and down the slopes, so that produce may be brought down the slope through the cleared coupe. All material should be removed before flushing begins, so as to avoid damage to the young shoots.

After repeated coppicing some stools become damaged during extraction of produce or by burning slash nearby and they die. Blanks caused this way are filled up with rooted cuttings or plants raised from seed at the same time as the coppice is cut. These plants should be 0.5–1.0 m tall and have strongly developed roots. In the case of *Azadirachta indica* and *Cassia siamea* coppice in the African savannas it is customary to replenish with stumped plants, that is, strong nursery plants with the stems cut off near ground level and the roots pruned. The length of stumped plants is generally 25 cm and 22.5 cm should be below the root collar; the stem diameter should be 1.5–2.5 cm. Another method that can be used for species with strong powers of vegetative reproduction is a method of layering called 'plashing' which was used in the southern parts of England to fill blanks in lime, ash, and Sweet chestnut coppice. At the time the coppice is cut a certain number of the smaller shoots are cut half through near ground level, bent over, pegged down, and covered with earth and sods. The production of roots and shoots is stimulated by making longitudinal cuts at intervals through the bark or by twisting the shoot so as to wound it. If new roots and shoots are successfully produced the layered shoot is severed from the stool after 2 or 3 years. 'Plashing' should be done in the dormant season.

Thinning coppice

Where coppice is worked intensively thinning is an important operation, the number of thinnings done depending chiefly on the length of the rotation, the competition among the shoots, and the markets for the produce. In coppice worked on a short rotation thinnings are usually dispensed with, although cleanings are sometimes done to remove unwanted species. Thinnings have for their object the improvement of the quality of the final crop by removal of congested or defective shoots; this work also induces more rapid growth in the shoots retained in the crop. Thinnings also furnish produce of different sizes and may increase the financial returns.

It is possible that there is, per clump, a definite number of shoots corresponding to a maximum production of volume and money; though the actual number in different cases may be expected to vary with species and local conditions. Howland (1969) measured *Eucalyptus* coppice grown for fuel at Muguga in Kenya and found that removing all but two or three shoots per stool at 18 months gave the highest yield in volume and value at 8 years. In other instances the beneficial results of thinning do not appear to be so marked; measurements recorded by Edie (1916) on teak coppice in Maharashtra state in India showed that thinning had little effect on the subsequent growth of the coppice.

Special forms of coppice

There are certain forms and modifications of coppice which deserve special
. mention. The most important are:

basket willow coppice;
short rotation coppice for energy;
coppice with field crops;
Eucalyptus coppice; and
pollarding.

Basket willow coppice

The coppice system on a short rotation is well-illustrated by willows grown
for basket-making. The cultivation of osier beds or 'holts' is done in river
valleys and alluvial plains in Europe, North America, South America, and
elsewhere.

The species most commonly grown are *Salix triandra*, *S. purpurea*, and *S.
viminalis* although *S. alba* and hybrids of *S. rigida* and *S. gracilis* called *Salix
× americana* are also cultivated. The best quality baskets are made from
1-year-old shoots from which the bark has been removed. Bark may be
peeled off in the spring, leaving a pure 'white' rod. If the bark is softened by
boiling to facilitate its removal, the decomposition products of the bark stain
the rod a pinkish-buff colour. 'Buff' rods are cheaper and used for general
purposes. Rods with the bark left intact are 'brown'. *S. triandra* is the most
suitable species for good quality basketry because it produces very supple
rods of the required dimensions and can be processed into equally high
quality 'white' or 'buff' rods. 'Black maul', 'Champion', and 'Newkind' are
among the preferred cultivars in England.

Salix purpurea produces adequately supple, smaller and slender rods but
does not process into good 'buff'. Cultivars of this species used to make
fancy baskets of white willow rods are 'Lancashire Dicks', 'Leicestershire
Dicks', and 'Dicky Meadows'. Rough agricultural and fishing baskets are
made from unpeeled 'brown' rods. The vigorous osier *S. viminalis* is used
widely in Europe. It produces rods of a quality usually adequate for agricultural
and fishing baskets, though some fishermen prefer to use 'brown' rods of the
more vigorous cultivars of *S. triandra* like 'Black Spaniard', particularly for
lobster pots. For the roughest work, orange-stemmed cultivars of *S. alba* such
as *S. alba vitellina* are often used. The American osier *S. × americana* is grown
in Germany, Poland, United States of America, and South America.

The cultivation of basket willows is an agricultural rather than a forest industry. The soils on which they are grown range from sandy loams to clays, the more moist and heavier soils producing rods of better quality than the lighter soils. Undrained, waterlogged soils are not suitable. In preparing land for willow cultivation the practice is to break up grassland in autumn to a depth of not less than 25 cm, leave it fallow during the next summer and plough it in the succeeding autumn; cuttings are then planted into weed-free soil during the following spring. Drainage is by open ditches, usually 60 cm wide and 45 cm deep, the distance apart varying with soil texture and the intensity of drainage required.

'Sets' or cuttings 23–38 cm long are taken from the large ends of selected 1-year-old rods, two to four sets being cut from each rod. Firmness in the soil is important and although sets 23–30 cm long will suffice on clay soils longer cuttings are needed on softer, peaty soils. The sets are pushed into the soil by hand or machine so that two-thirds are below ground and the remaining 7 cm, with at least two buds, protrudes. In southern England land ploughed in autumn is planted in March or early April.

Close spacing is general as this produces straight slender rods, reduces the development of side shoots, and suppresses weeds. The spacing actually adopted depends on the machines used for cultivation, spraying herbicides and other chemicals, and harvesting. In England (E. M. Liddon and N. Hector, personal communication) it is 60 cm between rows and 13–15 cm in the rows; in Hungary (Tompa 1963) it is 70 cm between rows and 15–30 cm in the rows, depending on the type of rod required.

During the first two seasons after planting, basket willow beds need frequent weeding; mechanical cultivators and chemical herbicides are used in a planned sequence. Thereafter frequency of weeding can be reduced, but on fertile alluvial soils best suited to basket willows, grasses, docks (*Rumex* spp.), Stinging nettle (*Urtica dioica*), and Creeping thistle (*Circium arvense*) soon form a dense cover if not controlled. Bindweed (*Convolvulus arvensis*) is probably the most troublesome because it runs from rod to rod. The addition of fertilizer is beneficial to the yield of rods on certain soils. Tompa (1963) prescribed nitrogen, phosphorus, and potassium on mineral and peaty soils in Hungary.

Protection of the tips of developing shoots from damage by fungi and insects is essential because any check or dieback reduces the length of the shoots and stimulates side shoots to develop so making the rods useless for basket making. The value of the rods is also reduced by lesions on the shoots which appear as disfiguring marks and points of weakness on peeled rods. The growing tips and young leaves are delicate and easily damaged by frost, wind, and hail.

Black canker, *Physalospora (Glomerella) miyabeana*, is the fungus most commonly concerned in damage to basket willow rods. The first symptoms are usually observed towards the end of May in Britain (Peace 1962). The infection starts in the leaves and passes to the shoots causing small lesions which eventually develop into cankers. This disease can be particularly serious in *Salix triandra, S. alba vitellina,* and *S. × americana.* The Willow scab, *Fusicladium (Pollaccia) saliciperdum* also attacks the stems and leaves causing small lesions; there can be considerable loss of foliage accompanied by dieback of the rods. *Salix alba* and *S. alba vitellina* are susceptible to this disease. Control of Black canker and Willow scab is possible by spraying with fungicide at intervals of 2 to 3 weeks, commencing just before the stools flush. In addition, the stools are cut low so that the fungi cannot overwinter on the snags. Rejected rods should not be left lying near the beds lest they serve as centres of infection (Peace 1962).

The rust fungi *Melampsora amygdalinae* and *M. allii-salicis-albae* can cause considerable defoliation and also attack the young shoots, producing small black cankers. The lesions reduce shoot growth and appear later as scars on peeled rods spoiling them for basket-making. Some control of infection by these rust fungi can be gained by late cutting of the rods in spring or removing the first crop of new young shoots when they are a few centimetres tall; in practice this can be done by grazing the beds with cattle (Stott 1956; Peace 1962).

A troublesome insect pest is the Brown willow beetle, *Galierucella lineola.* The adult beetles eat the growing tips of the rods early in July and the larvae skeletonize the leaves (Bevan 1987, p. 45). The combined damage stimulates side shoots to form thus spoiling the rods. Another defoliator is *Earis chlorana* which is common in Europe and North Africa. The larvae live among the leaves rolling them up with silken thread so reducing the growth of shoots and causing side branches to develop. The adults of the Osier weevil, *Cryptorhynchus lapathi,* insert their probosci into developing rods so weakening them that breakage can occur at the puncture; cork is also formed around the wound which causes unsightly scars when the rod is peeled. The larvae feed inside the stools and are often a major cause of the deterioration of basket willow beds (Stott 1956). Control is by spraying insecticide (FAO 1980) and careful cutting of rods and stools. The Button top midge, *Rhabdophaga heterobia,* lays its eggs on terminal parts of rods and the larvae, when feeding, cause a rosette of leaves to form which stops further growth (Stott 1956); side shoots develop and the rods are rendered useless. The sap-sucking aphids *Aphis farinosa* and *Cavariella pastinaceae* feed on the top of the shoots and in some seasons practically every willow bed in Britain is infested.

The holt should be coppiced in the first year after it is formed, although the crop is of little value, the shoots being often branchy and crooked. Thereafter cutting is done every year. The first saleable crop comes in the third season and increases in quality and value till the seventh year when full yields are produced. The season of cutting for 'buff' or 'brown' rods extends throughout the winter and into early spring. If 'white' rods are wanted, cutting is delayed until March when leafing-out is imminent, or more commonly, graded bundles of freshly cut rods are stood in about 15 cm of water, either in ditches or concrete pits. When leaves appear the rods are easily peeled. In England cutting was done by hand; now harvesting machines are available.

Although annual coppicing is the rule, some growers give a rest every few years. In Poland, Jezewski (1959) reported that a rest every 6 years resulted in greater weight of roots, better development of root systems, and higher yield of rods. If the area is divided into, say, five equal blocks, one block can be rested in turn, so the actual area being cut annually will be four-fifths of the total, three-fifths consisting of shoots 1 year old and one-fifth of shoots 2 years old. The latter are too large for normal basket work but are used for basket frames, strong hampers, and hurdles.

A well-tended basket willow holt should last 20 years; there are many that have been cropped for 50 years and a few for 70 years. The annual yield from the third year varies with locality, species, and management from 10 to 20 t/ha, a good yield being 15 t. Beds that are no longer profitable are grubbed up and laid down to grass for several years.

Osier willows have been grown from very early times. The first Roman writer on agriculture whose book has survived was Cato (234–149 BC) and, when listing the various crops in order of their profitability (Meiggs 1982), he placed willow beds third (below vines and the irrigated garden) and coppice seventh, above the orchard and mast wood. At the time of the Roman invasion the Britons were already skilled basket makers, using wickerwork for making shields and coracles or light boats. During the nineteenth century in Britain, the demand for containers was strong and willow-growing flourished, but the area of basket willow coppice has fallen during the past 60 years. The holts are now concentrated on the Somerset levels around Langport in southern England. The main areas in Europe are in Yugoslavia, Hungary, and Rumania on the flood plains of the river Danube and its tributaries. It is probable that there are 100 000 ha of Osier willows in eastern Europe (FAO 1980). Another important area of willow-growing is the delta of the Parana river in Argentina.

Short rotation coppice for energy

During the past 20 years, because of uncertainty about the exploitable reserves of oil and gas remaining in the world, the coppice system has been re-examined as a possible means of producing energy from biomass. The results of ecological, genetic, physiological, and silvicultural research have been applied in trial coppice plantations, and mean annual increments of dry stems and branches with bark of 10–12 t/ha per year have been .obtained over periods of 4 to 5 years (Cannell 1980).

The produce has been harvested by partly or fully mechanized methods and used directly as fuel or as a feed stock for processes that convert the cellulose to liquid fuel or gas. So far few large areas of short rotation coppice for biomass and energy have been planted but the system has attracted continued interest. Methods of producing suitable forms of forest biomass for energy are being worked out in many countries (Mitchell and Puccioni-Agnoletti 1983). Research on the sites and their treatment, the trees and their treatment, the harvesting of produce, and its subsequent conversion to energy is being sponsored and co-ordinated by the International Energy Agency. Because the end product is energy, some workers consider that the energy expended throughout the production process must be set against the energy finally produced when judging the viability of the coppice system.

Several broad-leaved species in the temperate zones are being tested in Europe and North America. The characters being sought include rapid growth in the juvenile stage, high yields of dry matter per unit of land, water, and nutrients provided, and suitable habit of growth—which is usually considered to consist of erect stems with compact crowns. Resistance to damaging pests and diseases and tolerance of frost and drought are essential characters. The wood properties required for energy are high specific gravity, low moisture content, and thin bark. Species that satisfy these criteria must also be easy to raise from seed or cuttings, establish themselves quickly after planting and have high coppicing or suckering ability.

So far species of *Alnus*, *Platanus*, *Populus*, and *Salix* appear to be the most promising and it is probable that species of *Acer*, *Nothofagus*, *Betula*, and *Fraxinus* together with Sweet gum, Yellow poplar and Locust tree will also prove suitable for short rotation coppice grown for energy. Most of the successful species are pioneers which occur early in the ecological succession. They usually bear large leaves which can achieve high rates of photosynthesis and produce a high leaf area index. Genetic variation is high among these species, and selection, progeny testing, and vegetative propagation have been used to produce cultivars that can respond to the intensive cultural treatments which are applied.

Because rotations are short the first crop of maiden trees to be coppiced must be established quickly. The site is completely cultivated and a period of fallow with repeated discing and chemical control of weeds prior to planting benefits the growth of the trees. The planting stock consists of carefully graded seedlings or cuttings, either bare-rooted or in containers. Planting usually is done at close spacing by machine, the actual distance depending on the size of produce required in relation to the method of harvesting and subsequent processing. When light mechanical harvesters are to be used several rods or poles of small diameter are normally required; if cutting is to be done by chain saw, fewer shoots of larger diameter are preferred. Although initial spacings of 0.3–1.5 m have been tried, planting distances of 1.2 × 0.8 m, 1.2 × 1.0 m, and 1.2 × 1.2 m are now becoming common; the final yields at 4 or 5 years are not much affected by initial spacings within this range. The number of plants lies between 7000 and 11 000/ha.

During the early years after planting and until a full canopy of leaves is produced the trees are dependent on supplies of nutrients from the soil, because the amounts cycling through the trees are inadequate for their needs, and competition from weeds can be severe. Thus, for the first 2 years after planting complete control of weeds is essential. This can be done by hand tools, machines, or chemical herbicides and all possible combinations of these methods are being tested. If additional nutrition is needed it must be applied in the early years before closure of the canopy. Miller (1983) has prepared schedules for applying herbicides, and nitrogen, phosphorus and potassium on a range of soil types and has suggested that regular monitoring of the nutrient content of the foliage is a possible basis for timing the treatments.

Irrigation is also being applied to short rotation coppice for energy where the potential water deficit exceeds 200 to 300 mm during the growing season. The methods being tested include flooding, delivering water to furrows between the rows of trees, broadcasting water over the canopy by travelling sprinkler systems, and providing it through continuous drip lines at the surface of the soil. Cultivars of *Populus* are more sensitive to water than many other broad-leaved species and require high soil moisture to maintain fast rates of growth.

The total input of energy used when compared to the energy value of the out-turn has been estimated by Zavitkowski (1979) as about 1:14 before harvesting begins, 1:9 after harvesting and chipping, and 1:4 as dry chips.

At present the advantages and disadvantages of short rotation coppice for energy appear to be as follows. The advantages are:

1. It is simple in operation and can be applied on small areas of under-used land as well as on larger areas.

2. It is a flexible system. If an existing species or cultivar is surpassed in health, yield, and quality of produce by another it can be replaced. Improved cultural techniques can be introduced quite quickly and, within limits, the size of produce can be adjusted to suit changes in
   · harvesting and processing.
3. Because of the short rotation only small amounts of capital are tied up in the growing stock and early returns are obtained.

The disadvantages are:

1. Financial success depends on the costs per unit of energy being lower than that derived directly from coal, gas, oil, nuclear power, wind, and other sources. The alternative markets for the produce are pulp, panel products, and fodder for animals which all require bulk supplies of raw material at low cost.
2. Very rapid growth is required which normally can only occur on fertile land with good supplies of water during the growing season. Nutrients may have to be added on many sites and irrigation may also be required on some sites.
3. Fencing is essential to protect the young coppice shoots from browsing by wild and farm animals. The rapidly growing shoots and foliage will also require protection from disease and insects which reduce yield and must be controlled.
4. Full mechanization of cultural operations and harvesting will require even terrain and firm soils.

Thus short rotation coppice for energy is akin to an agricultural system in its site requirements and intensive culture of a large number of plants. It is likely to be most suitable in areas where land and labour are under-used and other forms of energy are expensive. The needs of communities in Africa, Asia, and Central and South America which are almost wholly dependent on timber for fuel and many other purposes are discussed on page 211.

Coppice with field crops

Fr. Sartage, taillis sorté; Ger. Hackwald

An old practice extensively followed in certain parts of Europe, particularly in Germany, was that of coppice combined with the temporary cultivation of field crops. It is described here because of its similarity to agro-forestry practices in many tropical countries where the land supports both forest trees and field crops for a short period (page 240).
   Economically the practice is an interesting one. In Germany it was

confined for the most part to tracts situated in somewhat remote hilly country such as the Odenwald in Hessen, where the climate is severe and the soil poor. Under these conditions agriculture was difficult if not impossible without the aid of forestry. As a means of supplementing the local food in remote districts where the proportion of land under forest is large in comparison to that under agriculture, the practice of *hackwald* proved a great boon to the population.

The coppice consisted as a rule of oak for the production of tan-bark although occasionally there was a mixture of other species, such as ash, maple, hazel, hornbeam, birch, etc. The field crop most commonly grown was rye, although wheat, oats, and potatoes were also cultivated. The procedure varied in detail. In the Odenwald it was usual to work the oak coppice on a rotation of 15 to 20 years. The coppice was cut in May, as soon as the bark could be easily stripped. After the utilizable material had been removed and stacked the slash remaining was spread over the coupe and thoroughly burnt as soon as it was sufficiently dry, after which the whole area was hoed up by hand; a second hoeing was done in October, and the area was then sown with cereals. The crop was reaped with a sickle during the following June or August after which the area was abandoned to the coppice regrowth. If potatoes were grown they were generally cultivated for two successive years before the area was abandoned. It is obvious that the laborious process of hoeing and reaping by hand would not be practicable except in regions where necessity compelled the local people to resort to such methods. The burning, while supplying manurial ashes to the soil, did not appear to damage the stools to any extent; it had the effect of causing sprouting low down on the stool, which is advantageous. The stools themselves lasted many rotations, after which they were replaced by young oak plants with stems cut off near ground level.

In the Odenwald there was striking evidence that the practice of *hackwald*, far from exhausting the soil, created favourable soil conditions. It is remarkable that after the periodic burning of the area and the cultivation of the soil, the ground became covered with broom (*Sarothamnus scoparius*), heather (*Calluna vulgaris*), being rare or absent; on the other hand, where the coppice was worked without intermittent burning and cultivation, broom was generally absent and the ground became covered with heather, indicating acid soil conditions.

In Uganda *Eucalyptus* coppice has been successfully combined with field crops. Among the field crops grown with coppice are cereals, vegetables, and fruits. As is shown in Chapter 20, the possibilities of combining trees with food crops and animals are being thoroughly explored by the International Council for Research in Agroforestry (ICRAF).

Eucalyptus coppice

Most of the 4 million ha of *Eucalyptus* crops in the world are believed to be managed as coppice. Métro (1955), Jacobs (1980), and Evans (1982) have assembled information about how these are treated.

In species of *Eucalyptus*, coppice shoots develop from dormant buds situated in the inner or live bark or from lignotuberous buds found near the junction of root and stem. The lignotubers appear in seedlings as small protuberances in the axils of the cotyledons and sometimes of the first pair of leaves. They coalesce around the stem and fold downwards over the junction of stem and root to become wholly or partly buried in the upper layers of the soil. Lignotubers are storage organs which have the potential to produce leafy shoots in profusion if the above-ground parts of the tree are destroyed. A few species, which include *E. regnans, E. grandis, E. gomphocephala*, and the southern temperate form of *E. camaldulensis*, do not form lignotubers; these develop a thickened zone, like a carrot, at the junction of stem and root which also serves as a storage organ and new shoots develop from the top of it if the stem or crown are destroyed.

Eucalyptus coppice is normally cut at a maximum height of 12 cm, the stool being given a sloping surface to prevent water from settling. Chainsaws and bow saws are used for the cutting and as smooth a cut surface as possible is made; axes are not recommended because there is a higher probability of loosening the bark on the stool. Close supervision of the workers to ensure consistently low height of cutting is essential for the production of stable coppice shoots. This is because the stems of *Eucalyptus* species carry numerous dormant buds along every decimetre of their length and shoots will develop from all of these when the tree is felled. The higher shoots tend to develop faster than the lower ones and soon suppress them, but the upper shoots are much less stable because the callus which develops at heights greater than 12 cm up the stem is weaker and cannot give such strong support to the new stems as callus developing from a low cut.

If the produce taken from a coppice coupe must be peeled, the peeling or stripping of the bark is often done before the maiden trees or coppice shoots are felled. A light cut is made around each stem at a height of about 25 cm and the bark is stripped upwards from this cut; the bark can sometimes be pulled off to a height of 20 m! The maiden trees or coppice shoots are then cut at the maximum height of 12 cm.

After cutting, the slash is removed from the stools so that the coppice shoots may develop without interference. At first a great number of shoots grow from the stools but they gradually thin themselves out. However, it is not always the vigorous shoots that become permanent. The shoots crowd together in 'epicormic knobs' (Jacobs 1980), and the larger ones frequently fall or are blown outwards. In vigorous coppice this process may occur two or three times in as many weeks, but finally from two to six shoots remain firmly attached to the stool. It is from these that the next crop of coppice shoots is derived.

Fortunately most species of *Eucalyptus* are adaptable and generous in their response to season of cutting, but very dry periods and severe frosts can loosen the bark after cutting and cause the cortex to separate from the wood. Thus if the climate has a dry season the coppice should preferably be cut from just before the rains begin until the middle of the rainy season. In regions where frosts occur early in the growing season the coppice is normally cut when the incidence of severe frosts has decreased, but not so late as to impair resistance to cold at the onset of the next winter.

A wide range of produce is obtained from *Eucalyptus* coppice, but the most common is small roundwood for making pulp, paper, chipboard and fibreboard, timber for deep mines, a variety of posts and poles, and fuelwood for homes and industries, including charcoal for smelting iron in Brazil.

In most *Eucalyptus* coppice the first crop of maiden trees is felled at between 7 and 10 years. A common sequence was a total of four crops in 22 years with cuttings at 7, 12, 17, and 22 years but it has often been found advisable to omit the third crop of coppice, mainly because of poor finishing of the stools when the coppice is cut. Loss of stumps rather than loss of vigour in live stumps is a more common cause of reduced mean annual increment during several coppice rotations. In Israel five successive coppice rotations of *E. camaldulensis* have been obtained and the *E. globulus* coppice plantations of the Nilgiri hills in southern India have been cut on a rotation of 10 years for almost 100 years and still produce very good yields of fuelwood. At least two coppice crops after the original crop of maiden trees can be expected on short rotations of up to 10 or 12 years. This applies to *E. camaldulensis, E. cloeziana, E. globulus, E. grandis, E. maculata, E. paniculata, E. saligna*, and *E. tereticornis. Eucalyptus* coppice has been established under a very wide range of site conditions, species, and cultural treatments so that statements of mean annual yields of dry matter, are not useful unless they are accompanied by descriptions of each stand. Thus we must content ourselves with some general indications of yields per hectare per year. They range from 5 to 10 m^3 on severe sites, to 20–30 m on many favourable sites, and reach 40–50 m^3 where the combination of site, species, and cultural conditions are exceptional (Cannell 1979; National Academy of Sciences 1980).

A plantation intended to be managed under the coppice system requires:

(1) a soil with satisfactory nutrient status and supplies of water;
(2) good seedling stock of a suitable species and provenance;
(3) sound planting methods; and
(4) effective control of weeds.

The aim should be to enable the planted trees rapidly to colonize the site and attain uniformity of growth in height and diameter. There will always be some variation, but the more uniform the crop is and the smaller the range of diameters at the base of the stems the better is the survival of the stools and the higher is the productivity of the coppice. Irregularity in the maiden crop tends to become progressively greater in the first and subsequent coppice crops.

Many countries are conducting research into species and provenances of *Eucalyptus* and the results of species and provenance trials are being applied more widely, often through international co-operation in seed collection and distribution stimulated by the Food and Agriculture Organization of the United Nations (FAO) and the International Union of Forestry Research Organizations (IUFRO). It is, therefore, generally possible to select a species and provenance likely to be well-adapted to the site and capable of producing vigorous and healthy crops with good growth habit.

The irrigated plantations of the Gezira in Sudan provide an example of the successful adaptation of the coppice system to difficult conditions (Laurie 1974). The Gezira is primarily a cotton-growing area but the people employed in growing cotton need fuelwood and poles. The soils are alkaline black clays which crack in the severe droughts of summer. Rainfall is about 400 mm per year and the climate is classified as semi-desert, but irrigation allows the creation and maintenance of extensive coppice crops of *E. microtheca* which tolerates the heavy alkaline soil and the dry period of 3½ months when irrigation is not available.

The present working plan for Gezira prescribes an initial 8-year rotation of seedling maidens followed by 6-year coppice rotations. The planting stock is raised by direct seeding into polythene pots and is ready for planting after 5 to 6 months. The site is ploughed to form a series of low ridges, separated by the irrigation channels, and from 2.4 to 2.7 m apart. The plants are spaced 2.5 m apart on the sides of the ridges just above the irrigation water level. They are irrigated at intervals of 14 days during July to the middle of March, with a pause in October when cotton has priority and water is short. No irrigation water is available from the middle of March to the end of June because of the Nile Waters Agreement between Sudan and Egypt. The irrigation adds the equivalent of 640 mm of rainfall each year. Recently dieback appeared in some of the older plantations and the cause has been traced to insufficient water.

The original provenance of *E. microtheca* produced trees of poor stem form in the maiden crop, but much straighter poles have been obtained from the coppice crops of subsequent rotations. Trials of other provenances of *E. microtheca* and other species are being made mainly to obtain better stem form. The yields of timber from the irrigated crops average 60 m³/ha at the rotation age of 8 to 10 years, but growth is very variable. Improvements in site preparation, planting, and tending are being made to achieve more uniform crops with higher yields of produce.

The methods for dealing with slash left on the ground after the first and subsequent cuttings in *Eucalyptus* coppice vary. If the slash is spread irregularly over the area it can impede access and become a severe fire hazard; if it is destroyed by broadcast burning the resulting fire may kill a substantial proportion of the stools required for the next crop of coppice shoots. In Zambia, however, controlled burning very early in the rainy season is an essential element in the management of coppice. Sometimes the slash is placed between every third row of stools where it can be burnt on damp or windless days or left to decay. Retaining the slash protects the soil to some extent and reduces weed growth.

Observations at Natal in South Africa on a first cut of *E. grandis* at 7 years of age showed that stools between 10 and 20 cm in diameter survived cutting well and had low mortality; smaller stools of 3 to 10 cm in diameter and very large ones of 20 to 35 cm diameter showed high mortality after cutting. Experience in South Africa also suggests that the natural mortality in *E. grandis* coppice averages between 3 and 5 per cent each year and the initial planting distance and subsequent filling of gaps must take account of this.

Cleaning a crop of maiden trees before the first cutting is rarely necessary, but it may contain individuals of unwanted species or hybrids or with very defective stems and all these should be removed. Sometimes a young tree is damaged by animals or storms. If such trees are cut close to the ground a strong new shoot will develop. This early coppicing of damaged trees may have to be extended over all or a large part of a crop that has been damaged by fire, snow or wind so as to produce a sufficiently uniform stand more quickly than by replanting.

If five or six coppice shoots are left on a stool they will become bowed and suitable only for produce of low value. If straighter, more valuable produce is required the coppice must be thinned to three, two, or even one shoot. The sooner the final number of shoots per stool is attained the larger they will be at the end of the rotation, but too early reduction may cause loss of produce that might be sold after another year of growth. Very small but straight *Eucalyptus* stems are used as fence droppers in many places.

. The shoots left on the stool after the natural competition described on page 203 should have good stem form, wind firmness, and vigour before they are finally selected. Shoots that have come from dormant buds below the top of the stool and are firmly gripped by the callus should be chosen. Shoots growing on the windward side of the stools are less liable to windthrow than those growing on the lee side. Thinning coppice shoots should be done with light axes, bill hooks or machetes because chain-saws cannot be used among the congested shoots on the stools. The final number of coppice shoots after thinning should not be less than the original stocking. If stools have died, adjacent stools should be left with more than one stem.

Normally the yield from the first coppice crop is higher than that of the maiden seedling crop; provided that the cutting is done carefully, the species is one that coppices well, and an average of two to three stems per stool is retained to compensate for stools that fail to produce shoots, then the yield of the first coppice crop will be at least 25 per cent more than that of the maiden crop. There will be a gradual decline in yield from the second and subsequent coppicing. In India where *E. globulus* was coppiced four times after the maiden seedling crop, on a rotation of 15 years, a fall in yield of 9 per cent in the third coppice rotation, and of 20 per cent in the fourth and final rotation was recorded. The same species is grown in northern Spain where it is coppiced after 12 years and again at 24 and 36 years. Production falls off after each coppicing and it is recommended that the stools should be replaced after 36 years (Rouse 1984).

If an old crop of *Eucalyptus* coppice is to be replaced the stools should be removed or killed because shoots from them will interfere with the new crop, thus spoiling its uniformity; the old stools will also harbour pathogens which will attack the new crop. If there are 1000 or more stools per hectare in the old crop the stumps and large roots contain a considerable volume of wood that will make excellent fuel, and if the local demand is strong enough removal and stacking of this material will be justified. If the stumps cannot be removed they must be killed by destroying the dormant buds above ground and lignotubers below ground. This can be done with axes and chisel bars or by poisoning. Another method is to break off shoots produced by the old stools when weeding the new crop until it overtops the old coppice and the weeds.

Pollarding

Fr. Taillis sur tétards; Ger. Kopfholzbetrieb; Span. Tratamiento por trasmachos

Under ordinary conditions the practice known as pollarding can hardly be termed a forest operation, but it can be a component of an agro-forestry system (see page 240). It consists of cutting the tops off trees so as to stimulate production of numerous straight shoots on the top of the cut stem. These are trimmed off periodically at intervals of one or more years to provide material for basket-making, fencing hurdles, fascines, etc.

Pollarding of poplars and willows is commonly done along the sides of rivers, streams, and ditches. It was frequent in Europe around the margins of moist meadows where the pollarding was done at heights of 2.5 to 3 m— well out of the reach of cattle. Now the old pollards, often hollow and harbouring all the possible enemies of poplars and willows are everywhere disappearing from the landscape (FAO 1980). By contrast, pollarding is common throughout the dry regions of the tropics, in countries such as Ethiopia, Sudan, and Pakistan.

A similar operation is sometimes done on trees around fields and hop gardens to improve the shelter they provide by stimulating new branches along the stem. Another practice of a similar kind is the periodical trimming of the shoots that are produced along the stems of street and roadside trees to provide fuelwood. In China, London plane, certain poplars and other broad-leaved species are treated in this way. Pollarding is also done on farms in northern Nigeria where *Azadirachta indica* trees are grown for 12 years or so and then are pollarded and periodically lopped for fuelwood. About 30 trees treated in this way have supplied an average rural family's needs for fuel during 15 years (Fishwick 1965 in Spears 1983).

Advantages and disadvantages of the coppice system

Since the coppice system possesses special characters and presents a distinct contrast to the high forest systems, its advantages and disadvantages can best be appreciated by comparing it with high forest.

The advantages are:

1. It is very simple in application, and reproduction is usually more certain and cheaper than reproduction from seed.

2. In the earlier stages coppice growth is more rapid, and the poles produced are straighter and cleaner than in the same species when raised from seed. Hence where a large out-turn of poles or firewood billets of small to moderate size is required coppice is, generally speaking, superior to high forest.
3. Coppice is worked on a shorter rotation than most high forest crops. There is less capital tied up in the growing stock, and earlier returns are obtained than from high forest. Thus coppice is particularly suitable for small private properties in places where there is demand for the produce yielded by it.
4. The variety of habitat provided by different stages of worked coppice is beneficial to wild plants and animals, hence the conservation value of coppice is quite high.

The disadvantages are:

1. the out-turn from coppice consists of material of comparatively small size. There is therefore a limit to its general utility from the viewpoint of timber production, and its financial success depends on the existence of a special demand for the produce yielded by it.
2. Coppice draws heavily on the store of nutrients in the soil, particularly if the rotation is short, since it consists largely of vigorous young shoots and branches, which require more nutrients than older wood.
3. Young coppice shoots are particularly prone to damage from frost and from browsing by deer. Where frost damage is severe, or where deer are prevalent, the coppice system may have to be ruled out.
4. Aesthetically coppice is usually considered less desirable than most forms of high forest, since it is smaller and monotonous in appearance.

It is sometimes held that coppice is unsuitable for hillsides owing to the risk of erosion and soil deterioration due to periodic clearing of the area. The risk is probably less than is generally imagined, and is certainly less than in clear cutting in high forest (see O'Loughlin 1986). The stools of coppice on hillsides have strong root systems which serve to bind the soil, while the regrowth of coppice shoots quickly covers the ground and checks surface erosion. The protective value of coppice is well-illustrated in some of the Sweet chestnut coppice on steep hillsides in Italy.

Owing to its vigorous growth and the comparatively short rotation, coppice is usually considered to be less subject to attack by insects and fungi than high forest. In general this is probably true; but as Peace (1962) points out coppice stools almost invariably become infected with decay fungi. This form of attack can be delayed by cutting the stools low down to encourage production of shoots near or even below ground level.

Application in practice

Coppice is the oldest silvicultural system known. It has been traced back to Neolithic times in Britain and was used throughout the Bronze Age and during the Roman and Saxon periods. So far as we know it was the only form of forestry systematically practised by the early Greeks and Romans (Meiggs 1982). In Roman times coppice woods with annual coupes for the production of mine props, vine stakes, fuelwood, and other small material were termed *silvae caeduae*. Short rotations were adopted. Pliny, in his *Historia Naturalis* mentioned 7 years for Sweet chestnut for the production of vine stakes and 10 years for oak. In France and Germany coppice was practised during the Middle Ages, that is, from AD 337 to 1500, mainly for the production of fuel, short rotations being the rule. An early record of coppice in Germany is from the year 1359 in the town forest of Erfurt, and from the thirteenth and fourteenth centuries over large areas of Germany coppice was combined with the temporary cultivation of field crops. Rackham (1980) considered that by the year 1250 coppice was widespread in Britain, even in such large woodland areas as the Forest of Dean. In Switzerland, according to Flury (1914), coppice regeneration was first practised about the sixteenth century, although systematic cuttings were not done until the end of the eighteenth century. The Chinese fir has been worked on coppice rotations of 25 to 40 years to produce poles and firewood in the temperate parts of central south China, probably for hundreds of years (Lin 1956).

Throughout Europe, during the seventeenth and eighteenth centuries, coppice continued to supply domestic firewood, building material, fencing and, in addition, increasing quantities of fuel for industry and oak bark for tanning. Then in many parts of Europe coal began to supplement wood as a fuel and by the middle of the nineteenth century many of the traditional products of coppice were being superseded. In Britain this decline in the coppice system accelerated after the first world war as rural electrification spread (Crowther and Evans 1984), and again after the second world war as oil and gas became readily available. By the 1950s regular coppicing, apart from that of Sweet chestnut, had become rare and conversion of coppice to high forest was getting under way (Wood *et al.* 1967). On the continent of Europe, coppice had also become much less important than it was. Large areas of former oak coppice had already been converted to high forest and the conversion of much of the remainder was in progress. However, large areas still remain and although, for the reasons just given, the total area of coppice in Europe steadily diminishes, there are still instances where it gives remunerative results because there is special demand for the produce.

In France (Auclair 1982) there are still 5 million ha of coppice with an annual yield varying from 1.5 to 5 m^3/ha, depending on species and site conditions. Some of this coppice is stocked with commercially valuable species and the coppice stems have an average breast height diameter greater than 15 cm. According to Curdel (1973) and Bouvarel (1981) the production of these areas can be improved through enrichment of the stands, application of fertilizers to increase rate of growth, improving the network of roads to increase access, and harvesting the produce by machines. Fifty three per cent, or 3.6 million ha of the forests of Italy are coppice or coppice with standards and these have been deteriorating since the 1950s because of the decline in rural population, the rise in cost of labour, and falling demand for the produce. The best areas are being improved by similar methods to those used in France; attempts are also being made to overcome the problems caused by the small size of individual holdings by encouraging their owners to combine in working the coppice.

The census of woodlands and trees carried out in England from 1979 to 1982 (Forestry Commission 1983) revealed that the area of coppice was 25 711 ha; over half of this was composed of Sweet chestnut coppice which remains the most profitable form. Some 3000 ha of hazel coppice is worked on 7 to 9 year rotations with 1500 stools/ha for hurdles, hedging stakes, bean poles, and similar products (Crowther and Evans 1984).

The first kind of intentional silvicultural practice in the United States of America was the coppice system. Almost from the time of settlement until early in the present century accessible parts of the eastern broad-leaved forests were coppiced repeatedly to produce successive crops of fuelwood and charcoal for domestic and industrial use. Then the coppice system almost vanished, except for small quantities of specialized products or for producing browse for wildlife (Gysel 1957). During the past 20 years (McAlpine *et al.* 1966) it has been revived as a possible means of producing energy and chips for processing into pulp and particle board (see page 198).

The coppice system has been widely used throughout the developing countries and its importance is increasing as populations continue to rise. In these countries fuel wood is gathered from unmanaged and managed forests, from trees growing outside forests in rows and hedges bordering roads, canals and railways, from village woodlots and orchards, and from scattered trees in fields and around houses. The average demand for each person in Africa, Asia, and South America appears to be about 1 m^3/year and the total annual demand for domestic fuelwood is crudely estimated at 2000 million m^3 (Gilliusson 1985). As is now well known, supplies of domestic fuel are in severe deficit in the arid and semi-arid zones of Africa, the Indian sub-continent, and south-east Asia, and shortages are expected to spread to other parts of these regions.

The energy derived from burning wood is important in many rural industries. Gilliusson (1985) emphasized that wood can also be a major source of energy in generating electricity and providing a feed stock for liquid fuel used in vehicles. Thus two major uses of fuelwood in developing countries are to provide 'energy for survival' in the home and 'energy for development' in industry and transport. Long experience has demonstrated the value of the coppice system in providing sustained supplies of cheap fuelwood. Some species which coppice freely also yield fodder for animals, tannin, oils, and fruits. The shelter and site protection provided by managed coppice can greatly improve the quality of life for people in deprived areas (Burley and Plumptre 1985; Arnborg 1985).

17

The coppice selection system

Fr. Taillis fureté; Ger. Geplenter niederwald; Span. Tratamiento de monte bajo entresacado

General description

In principle the working of the coppice selection system is similar to that of the selection system in high forest. An exploitable diameter is fixed according to the size of material required, and an estimate made of the age at which material of this size is produced; this age determines the rotation, which is divided into a convenient number of felling cycles, and the area is divided into annual coupes equal in number to the number of years in the felling cycle. Each year coppice fellings are carried out in one of the annual coupes, but only shoots that have reached exploitable size are cut, those below this size being left.

In practice this system varies in detail. In the Pyrenees the most frequent rotation is 30 years with two felling cycles of 15 years or three of 10 years; a rotation of 27 years, with three felling cycles of 9 years, is adopted in some places. In the Morvan massif of southern France a rotation of 36 years, with four felling cycles of 9 years, is the general rule.

The clumps normally consist of shoots of two, three, or four different ages, according to the number of felling cycles in the rotation; these, however, are not always sharply distinguishable, the clump actually consisting of a few erect shoots of the larger dimensions and a number of smaller shoots, many of these being whippy, flexible, and sometimes even straggling.

At one time it was customary in some places to reserve a few of the larger shoots scattered over the coupe in the form of standards to give seed for the regeneration of blanks caused when the stools died off; this, however, has fallen out of use. Blanks are sometimes filled by pegging down whippy shoots from adjacent clumps and obtaining reproduction by layers. Beech stools are said to retain their vitality longer under the coppice selection system than under simple coppice.

Advantages and disadvantages

The coppice selection system is generally used on poor and often rocky ground in hilly country where the trees do not attain sufficient dimensions for high forest and under severe climatic conditions where young shoots worked under simple coppice would suffer from frost, drought, and snow pressure. Under these conditions the advantages are:

1. The young shoots are better protected from the dangers already mentioned as well as from grazing animals, by the cover of the older shoots.
2. The soil remains permanently covered and it is not subject to periodical exposure as in the case of simple coppice.

The disadvantages are:

1. Cutting of the large shoots is more difficult and tedious, and is apt to damage the smaller ones, while cutting at ground level is generally impossible.
2. Development of the shoots is poorer than in the case of simple coppice, owing to the suppression they undergo in the earlier part of their life.

Application in practice

The coppice selection system has long been applied to beech in certain parts of Europe, although it is now less common than it used to be. In France at the present time it is used in the Pyrenees, at elevations of between 600 and 900 m, and to a lesser extent in Haute-Savoie and the Morvan massif. It is common throughout the Balkan Peninsula. The sub-tropical broad-leaved forests of *Olea cuspidata* and *Acacia modesta* in the Jhelum Mianwali and Shahpur forest divisions of Pakistan are worked under the coppice selection system (Khattak 1976). The coppice selection working circle is divided into eight felling series. Yield is regulated by area, there being 30 annual coupes in each felling series. The exploitable size for *Olea* and *Acacia* is 15–20 cm diameter at 30 cm above ground level. Each year an annual coupe is gone over for shoots of exploitable size. Blanks are filled using dry zone afforestation methods.

Coppice selection forest rarely has a flourishing appearance, and it is not to be recommended except in the rather special conditions under which it is generally practised.

18

Coppice with standards

Fr. Taillis sous futaie; Ger. Mittelwald; Span. Cortas en monte bajo con resolvos; monte medio

General description

Coppice with standards consists of two distinct elements:

(1) a lower even-aged storey treated as coppice; and
(2) an upperstorey of standards forming an uneven-aged crop and treated as high forest.

The coppice is termed the underwood and the standards the overwood. The purpose of the standards is to provide a certain proportion of large timber, to provide seed for natural regeneration, and in some cases, to afford protection against frost. Coupes are formed exactly as in simple coppice. The rotation of the coppice is fixed according to requirements and the area is divided into as many annual coupes as there are years in the rotation. As each annual coupe in turn becomes due for felling, the following operations are carried out in it:

1. The coppice is clear cut as in simple coppice.
2. A certain number of the existing standards are reserved for at least one more coppice rotation and the rest are felled.
3. A certain number of new standards of the same age as the coppice, and preferably of seedling origin, are reserved. These new standards have arisen as natural seedlings appearing on the coupe or from plants introduced at the time the coppice is cut.
4. Blanks caused by death of stools or removal of standards are filled up to ensure a future supply of both coppice and standards. If natural seedlings are not present in sufficient quantity, plants of the desired species, provenance or cultivar are introduced by planting.

If these operations are repeated regularly for several coppice rotations of r years, then the coupe about to be felled should consist of coppice aged r

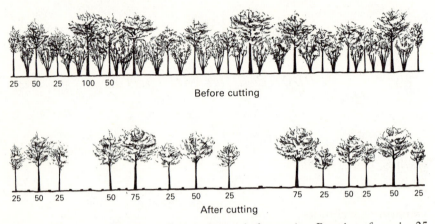

Before cutting

After cutting

Figure 30. Coppice with standards before and after cutting. Rotation of coppice 25 years; rotation of standards 100 years. Numbers denote age of standards.

years together with standards aged 2*r*, 3*r*, 4*r* . . . years, and a number of young prospective standards aged *r* years. In other words, the rotation of the standards is a multiple of the coppice rotation. If it is decided to retain standards for a maximum of four coppice rotations, then as soon as the felling has been done there will remain on the ground standards aged *r*, 2*r*, and 3*r* years, those 4*r* years old having been felled along with the coppice and those *r* years old having been newly reserved. This is shown diagrammatically in Fig. 30. The standards have shorter stems and a larger proportion of branchwood than trees grown in close-canopied high forest. This is particularly the case where the rotation of the coppice is short; with long coppice rotations the stems of the standards are usually free from side branches to a greater height.

The arrangement of the coupes to aid extraction is done as in simple coppice (see page 193). In Europe it was sometimes customary to arrange the coupe so that fellings proceeded in directions against cold winds so as to protect the young coppice.

Choice of species

The underwood in coppice with standards generally consists of a mixture of species. In Britain these may include oak, ash, hornbeam, beech, Sweet chestnut, sycamore, Field maple, hazel, alder, lime, elm, Wild cherry, sallow, and aspen, the last named regenerating from suckers. Areas of pure underwood, particularly of oak, ash, Sweet chestnut, and hazel also occur. Strong light-demanders are less suitable as underwood than species capable of standing some shade.

Standards should be of species sufficiently valuable to compensate for the loss of increment in the coppice. They consist, for preference, of species with open crowns carried some distance above the ground. It is, however, believed in Britain that some open-crowned trees, including ash and birch, make unsatisfactory standards because despite their thin crowns coppice grows poorly beneath them; this may be due to their dense rooting near to the soil surface. In Britain most of the older standards are oak, but during this century the neglect of sowing and planting has led to the replacement of much oak by standards of species that regenerate more profusely. Beech, although occasionally grown as a standard, is not well-suited mainly because of its densely foliaged crown. In Europe, oak is the commonest standard tree. Light-foliaged conifers, particularly the larches, make suitable standards (Köstler 1956, p. 367).

Classification of standards

The English and French terms commonly used are:

1r Teller   Baliveau
2r 2nd class standard   Moderne
3r 1st class standard   Ancien
4r Veteran   Bisancien
5r —   Vieille écorce

Standards of the various classes can generally be distinguished by their size (Fig. 31), particularly if the coppice rotation is a long one, but if the overwood is composed of several species with different rates of growth this may be difficult. Where there is more than one species in the overwood it is permissible to adopt different rotations for the different species, should their respective rates of growth or longevity demand it, or should the size of material required in the case of each species differ. Troup (1952, p. 147) quoted an example from the town forest of Rastatt in West Germany with a coppice rotation of 25 years. The rotations adopted for the standards were: oak, 125 years; ash, 100 years; birch, alder, and hornbeam, 75 years; poplar and Locust tree, 50 years.

Reserving and removing standards

The selection of standards for reservation should take account of:

(1) the average number of standards;
(2) their distribution over the area;

Figure 31.   Coppice with standards 1 year after cutting the coppice. This stand lacks sufficient tellers of oak. France.

(3)   the selection of straight, sound specimens of the best species with well-developed but not too spreading crowns; and

(4)   the maintenance of a correct proportion of classes.

It is usual to fix approximately the total number per hectare of standards of all classes to be reserved at the time the coppice is cut. More rarely the volume per hectare to be reserved is laid down, this varying with site capability; this is applicable only where the crop consists mainly of standards and approaches selection high forest in character. The number or volume of standards to be reserved depends on the relative importance of overwood and underwood. Every additional standard means a corresponding reduction in the out-turn of coppice; hence by fixing the number of standards an approximate ratio can be obtained between the out-turn of saw logs and that of smaller material. As a general rule between 50 and 100 standards/ha of all classes are reserved; with more than 100 the coppice, unless strongly shade-bearing, is relegated to a secondary position.

To ensure sustained yields, attention must be given to the correct age distribution of the standards. Just as in the selection system (see page 167), allowance must be made for mortality among the immature standards or for their removal on silvicultural grounds. Thus, for every 10 to 20 tellers reserved only 1 or 2 may survive to become standards of four rotations. The rate of mortality varies under different conditions and an average for any locality can be determined only by experience. As an example, assuming that 100 standards/ha are to be maintained, the proportion of these, immediately after the coppice is cut, would be somewhat as follows:

Class of standards	Tellers	Second class	First class	Veterans
Rotation	1	2	3	4
Number of standards/ha	50	30	20	(10)
	Just reserved			Just removed

Standards for reservation are marked at breast height and all trees not marked and coppice are felled. As a rule the standards are scattered singly over the area but sometimes they are concentrated in groups distributed among the coppice, or in belts with intervening belts of coppice, an arrangement that produces a larger proportion of clear timber on the standards. It is not always possible to find suitable standards distributed evenly over the area; the prescribed number per hectare should therefore be regarded as a general average rather than as a hard and fast figure applying to each hectare of forest. Some authorities have advised reserving more than the accepted number if good standards can be found. However, this ignores one of the main objects of the system which is to supply a certain proportion of small material; if the object is to produce as much large timber as possible then coppice with standards should be abandoned in favour of high forest. It is sometimes the custom to reserve a larger number of standards along the forest boundaries, as a protection against strong and dry winds, and near roads and rides so as to reduce the cost of extraction, and for amenity. Where frost damage is severe, it is sometimes the custom to reserve a large number of tellers to protect the young coppice shoots from damage; as soon as the risk is past these tellers are thinned out to the required number.

Standards should, if possible, be of seedling or clonal origin, since coppice shoots left standing for a long time tend to become unsound at the base. However, where there is a shortage of seedlings or cuttings, it may be necessary to reserve selected coppice shoots as standards. These should be chosen from young stools with few shoots on them. Suckers are preferable to coppice shoots, owing to the absence of wounds at the base.

The felling of standards is not confined to those that have reached exploitable age. It includes also the removal of younger standards which are dead, moribund, diseased or otherwise undesirable, and it may even extend to the removal of sound standards that are in excess of requirements. In exceptional cases standards that have reached the prescribed exploitable age may be retained for another coppice rotation, if they are of especially fine quality and are likely to increase materially in volume.

Standards should be felled and removed immediately after the coppice is cut, to avoid damaging young coppice shoots after sprouting takes place. If the standards are stripped for tan bark, this should be done during the spring before felling takes place, the barked trees being left standing until the following winter.

Tending operations

These comprise cleanings and thinnings and, where necessary, pruning standards. Early cleanings include removal of unwanted species, weeds, and climbers that are threatening coppice shoots and seedling trees, and also removal of coppice shoots that are interfering with seedling trees required for the future supply of standards; the latter is an important operation owing to the more rapid growth of coppice.

Thinnings among coppice shoots are done in the manner described for simple coppice (page 194). In addition, thinnings are made to free young prospective standards from coppice shoots that threaten them and to remove dead and dying standards and others whose removal may be desirable, provided this can be done without serious damage to the crop. The pruning of standards may also be done to produce clear timber. This operation includes removing epicormic branches which sometimes appear after the coppice is cut, and other small branches less than 5 cm in diameter; the removal of larger branches more than 5 cm in diameter should, in general, be avoided (Peace 1962). Where young standards are reserved in especially large numbers to protect the coppice against frost, these standards must be thinned out, when the coppice is free from danger, to the number ultimately required; this is done in one or more operations.

Coppice with coniferous standards

This consists of coppice with an open, even-aged overwood of conifers worked on the selection system with a felling cycle equal to the coppice rotation. It is seen in some of the mixed forests of Holm oak and Aleppo pine on the dry limestone hills of the Mediterranean coastal region of France (Troup 1952, p. 150). The oak is worked as coppice, the rotation being as a rule 25 years. The pine, which is tapped for resin, is worked to a minimum exploitable diameter limit which varies with the locality. The rotation is not fixed, but may vary from three to five coppice rotations. Natural regeneration of pine is relied on as far as possible, but supplementary direct seeding or planting is often necessary.

A more recent example of coppice with coniferous standards comes from the Republic of Korea (FAO 1979b) where the coppice is Locust tree and the standards are Pitch pine.

Advantages and disadvantages

The advantages are:

1. It provides material of different sizes in considerable variety and can supply a local demand for fencing material, pulpwood, fuelwood, charcoal, poles, turnery wood, and timber.
2. It provides early returns from the coppice, which is a financial advantage.
3. The forest capital is small compared with most forms of high forest, and it can readily be increased or decreased without affecting the system. This flexibility is an advantage for private owners.
4. Owing to the cover provided by the standards, the soil is better protected than in the case of simple coppice.
5. Replenishment of blanks is not done wholly by artificial regeneration, but is effected at least in part by natural regeneration, so reducing expenditure.
6. It is continually attractive in appearance, since clear fellings are avoided and the overwood consists of well-developed trees of different sizes.
7. The diversity of species and age classes makes it a good system for conserving wild animals and birds, especially because of the cover provided by the overwood.

The disadvantages are:

1. It is a difficult system to apply correctly. Maintaining the balance between coppice and standards and the correct distribution of standards of different classes is difficult. The selection of standards requires skill, and is tedious to do among a thick growth of coppice which impedes vision to any distance.

2. The standards are often more branchy and short-stemmed than trees grown in high forest and, in consequence, yield a smaller proportion of clear timber. The amount of small material, including branchwood, can amount to three-quarters of the total out-turn and much of it is fit only for fuel.
3. Coppice grown under standards is generally less vigorous than simple coppice.
4. Harvesting produce requires more labour than in equivalent separate areas of high forest and simple coppice, and this system has become unprofitable in many places. Harvesting by machines is more difficult to organize than in equivalent areas of high forest.
5. Concerning liability to damage, the coppice suffers more from browsing by deer when these animals are present. Older standards are windfirm (because they have developed as individuals) but young standards suddenly freed from the intervening coppice are liable to be bent or uprooted by wind and snow. Smooth-barked standards may suffer from sun-scorch when exposed.

Application in practice

Lanier (1986) summarized studies made in France to compare the out-turn in volume and value of simple coppice and coppice with standards with that yielded by broad-leaved and mixed coniferous and broad-leaved high forest. The data confirm the superior volume production of the high forest regimes but it also emphasized the greater variety of produce derived from high forest, which can supply a majority of industries based on wood. Table 2 summarizes the proportions of different classes of produce derived from four different regimes; it is based on a very comprehensive study.

Table 2. Summary of different assortments derived from forests
(per cent of total production)

Silvicultural regime or system	Waste and small wood	Fuel wood	Pulp and board wood	Sawlogs and veneer logs
Broad-leaved high forest	18	34	17	31
Coniferous high forest	13	14	25	48
Simple coppice	15	65	20	0
Coppice with standards	16	58	20	6

Source: Lanier (1986, p. 127).

In Europe it is generally considered that coppice with standards is inferior to high forest without possessing the advantages of simple coppice; if timber is required it should be grown in high forest while if fuelwood and poles are wanted simple coppice is preferable to coppice with standards.

In Britain until the middle of the nineteenth century, coppice with standards was the principal species applied to broad-leaves. When oak was in demand for shipbuilding and there was a good sale for firewood the system was highly profitable and was worked with close attention to the correct distribution of standards. With industrial development, the more extensive use of coal and the construction of railways, which distributed coal throughout the country, the demand for firewood diminished or ceased altogether. Furthermore, with the introduction of iron ships in the middle of the nineteenth century the demand for oak timber for shipbuilding declined rapidly. The importance of coppice with standards diminished and in many places there was little or no sale for the produce. Since the end of the second world war in 1945 a considerable area of derelict coppice with standards has been converted to high forest, but despite its disadvantages the system has continued to some extent on private estates partly in the form of wildlife reserves and partly to furnish supplies of timber, firewood, fencing, and other small material for use on the estates. There are still a few local industries dependent on this type of forest.

The latest census of woodlands (Forestry Commission 1983) revealed that only 11 500 ha of coppice with standards remain; most are in south-east and southern England and all are privately owned. There are also some 140 000 ha described as ancient and semi-natural woodlands by Steele and Peterken (1982) many of which were worked on the coppice with standards system and are now unproductive. These probably originated during the Middle Ages and they have existed continuously since that time. Proposals have been made to restore some of them to productive working so that the indigenous trees and shrubs and the associated flora and fauna may be conserved.

The coppice with standards system has also been used since the Middle Ages in France. It was practised in royal forests during the fourteenth century and reached its greatest extent during the eighteenth and early nineteenth centuries. In 1824 Bernhard Lorentz (born 1775, died 1865) began his experiments on methods of converting coppice and coppice with standards to high forest (Plaisance 1966). Fifty years later Tassy (1872) estimated the annual loss to France by reason of the large areas of coppice and coppice with standards at the very large sum of 293 million francs (roughly equivalent to £1 billion today). Enormous demands were made on the forest resources of France during the first world war, from 1914 to

1918, and the inadequacy of coppice with standards was again demonstrated. After the second world war a new surge of conversion began but France still has 3.9 million ha of coppice with standards, which are mainly situated in the central and eastern regions. Over 2 million ha are privately or communally owned as small stands often on steep slopes with difficult access. The policy in the national forests is to continue converting coppice with standards to high forest; this will be accomplished soon (Auclair 1982).

In Germany the system appears to have been used in some form since about AD 600. In those days it was combined with the pasturage of cattle and pigs, hence the standards consisted of food-yielding trees such as oak, beech, and fruit trees, while the underwood was cut periodically to provide fuel. This practice persisted throughout the Middle Ages and was still occasionally seen in the 1950s.

Since early in the nineteenth century the area under coppice with standards has diminished steadily, Saxony leading the way in conversions to Norway spruce. In the Federal Republic of Germany today there are 95 000 ha of coppice with standards, chiefly in communal forests of the west and south-west. Progress in converting this to high forest is slow because of the expense. Areas of ancient coppice with standards have been conserved. In Baden-Württemberg, for instance, there are three fine examples at Bechtalerwald in Oberrheinisches Tiefland, at Schnapsenreid in Odenwald, and at Stammberg in Neckarland (Dieterich *et al.* 1970).

The coppice with standards system is used in the tropics but there is little information about the areas involved. It appears to be widespread in India where there are about 1.6 million ha. It is important in that country for the local supply of firewood, house timbers, and other small material for village needs. It is much used on land that is not capable of producing timber of large size.

Although the demand for fuel, poles, and small timber is very strong in the tropics and the very wide range of species suggests that new combinations for the overwood and underwood might be found and developed, simple coppice and the clear cutting system using fast-growing species appear to hold out more promise in the long-term.

19

Conversion

Fr. Conversion, transformation; Ger. Überführung; Span. Conversion, transformación

Introduction

The term 'conversion' is generally taken to mean a change from one silvicultural system to another, as for instance, conversion of coppice with standards to high forest or from clear cutting to the group selection system (page 175). In silviculture, conversion also includes restoration of degraded forest to a productive condition, leading up eventually to sustained yield working under one of the silvicultural systems. The purpose of this chapter is to describe the methods used when making conversions and give examples of different kinds of conversions to supplement those already described in the chapters on the clear cutting and group selection systems (pages 80 and 175).

Natural indigenous forest, managed high forest, coppice with standards, and coppice can all become degraded. The most common causes are:

(1) uncontrolled exploitation fellings which remove the best trees of the principal species, leaving only damaged or moribund trees and secondary or weed species;
(2) changes in market or economic conditions leading to a decline in the value of the growing stock and neglect of the forest;
(3) severe damage by fire, wind, snow, disease, insects, or war;
(4) uncontrolled grazing by domestic stock or wild animals; and
(5) certain agricultural practices, particularly unregulated shifting cultivation and removal of litter.

There is a further element involved in many of these causes, namely that of ownership or rights of user (page 6). The owner or right holder in a given forest may be unable or unwilling to stop the practices leading to degrade and may need a strong incentive to make the forest productive again. The decision whether or not to convert a degraded forest to a productive condition or make a change from one silvicultural system to another is normally based on productive, protective, and social considerations. Conversion is appropriate where:

(1) the yield of the existing crop does not match the capability of the site for timber production and the owner seeks a higher income from the forest;
(2) the soils are liable to damage under the present silvicultural system or degraded forest and conversion to high forest is necessary to protect the site;
(3) markets and communications have improved and national policy favours conversion;
(4) there is need to provide employment in the forest and wood-using industries;
(5) the aesthetic quality of the forest must be improved;
(6) advances in silvicultural techniques make conversion practicable.

Conversion may not be appropriate when:

(1) the growing stock provides sufficient quantities of produce, such as fuelwood, to satisfy the needs of the owner or right holder and there are no incentives encouraging change;
(2) the existing forest occupies a particularly fragile site or one of low capability;
(3) it has conservation value as a refuge for wild plants and animals.
(4) it is suitable for field sports and recreational use;
(5) the aesthetic quality or historical significance of the existing forest is high.

The character of degraded forests are low stocking of the principal species and a relative abundance of secondary and weed species. Dense thickets and tight groups of pole-stage trees alternate with poorly stocked areas and unstocked gaps where the ground may be covered with strong growth of grass and other vegetation. The distribution of age and size classes of the principal species is very incomplete. All these characters make for irregularity in species, stocking, tree size, and tree quality which is often matched by soil variations, some due to damage done during extraction of logs.

If time does not enter the reckoning it is possible with modest and inexpensive silvicultural treatment to convert, say, coppice with standards to broad-leaved high forest; but nowadays it is not usually justifiable to hold land of good capability under poor forest while waiting for slow processes of ecological succession and competition to bring about improvement. In many cases the chances of obtaining a satisfactory crop by these means alone and within a few decades are small. Moreover if land is short the degraded forest could be taken over for agriculture or some other use.

The methods actually used in conversion, the amount of work that must be done, and its net cost depend on the present condition of the site and tree crop. The potential gains in volume and value of timber that are likely to accrue are determined by the capability of the site for forestry and the potential yield of the species, provenances, and cultivars used in the converted forest.

The first step in planning a conversion is to gather information about the site and growing stock. A survey of soil types and of the slope and roughness of the terrain is used to assess and classify the capability of the different sites. An inventory is made of the growing stock by species, age or size classes, top heights, current annual increment, stem form, branchiness, health, and stocking to assess its potential yield in quantity and quality of produce. As much information as possible is also gathered about past management of the site and growing stock so as to identify possible problems that may have to be faced in the future. Degraded forests often lack convenient access because they are surrounded by other, possibly conflicting forms of land use.

The second step is to examine the potential benefits of conversion—which include increased yields in volume of timber and higher proportions of valuable assortments, such as sawlogs and logs suitable for conversion into veneers. Thus in the case of a conversion of typical coppice with standards to high forest on sites of high capability for forestry in the lowlands of Britain and France, the current annual yield in volume and value of timber would be compared with the yield in volume and assortments of high forest stands of the same species, other indigenous broad-leaves and conifers, and exotics known to be suitable for the site. Because stocking can be improved and the rotation shortened by conversion, mean annual increment can readily be doubled or trebled with indigenous species, and further increased by the use of well-adapted exotics. The increased out-turn of sawlogs and veneer logs also improves the financial yields very considerably.

In the broad-leaved forests of Europe and North America, soil fertility is often high but the physical features of soil and terrain may have presented difficulties for land users in the past—which is why many sites are still under forest. Thus in the lowlands of England much degraded forest is on clay soils that are poorly aerated, with marked fluctuations in water content from too much in winter to too little in summer. When the existing forest is cleared the water table rises and more drainage becomes essential. If the terrain is flat, late spring frosts may become troublesome. The choice of species for the new crop is limited to slow-growing hardy species, and the potential returns from an investment in conversion are unpromising. The situation is better on the podsols and sands or gravels where the regrowth from cleared degraded broad-leaved forest soon declines in vigour. These soils are not suitable for broad-leaves and conversion to conifers is usually straightforward and inexpensive. Future yields of timber are predictable and the investment in conversion generally brings a good return.

The main silvicultural techniques used in conversion are:

improvement fellings;
enrichment, using lines, strips, or groups;
replacement following clear cutting or under a shelterwood.

Improvement fellings

Fr. Coupes d'amélioration; Ger. Pflegemassnehmen; Span. Cortas mejoración, cortas de mejoria

In many temperate and tropical countries there are very large areas of broad-leaved and coniferous forests which have been exploited for certain principal species and then left more or less neglected. As times passes it becomes evident that many of these degraded forests contain young trees of the principal species, but their growth is hampered by the presence of larger trees of lesser value which were not harvested originally. In the United States of America (Smith 1986) the growing stock of millions of hectares of exploited broad-leaved forest has been restored to a more productive condition by improvement fellings. The same applies over very large areas of India where improvement fellings are being used in the conversion of degraded forest to the tropical shelterwood system.

Improvement fellings are made in stands that have grown beyond the sapling stage to improve the species composition, growth increment, and quality of the growing stock by removing trees of unwanted species, slow growth, poor form, and unhealthy condition from the main canopy so that desirable trees of the principal species can become more productive. Those removed include secondary or weed species, trees with crooked or leaning stems, very branchy and badly formed trees, those that are over-mature and occupying much growing space, and those that have been seriously damaged by fire, insects, fungi, animals, and other agencies. Improvement fellings may be applied to stands of almost any combination of species, age, or canopy structure but there must be sufficient stems of the principal species that will respond to treatment and are considered to have suitable phenotypic or genotypic qualities.

Improvement fellings are most often used in conversion as a prelude to commencing a typical thinning regime in the improved stand; and as preparation for regeneration fellings in certain methods of conversion to another (see page 236).

If the defective trees have market value above the cost of felling they are harvested; if not they are killed by girdling plus herbicide. No useful purpose is served by restricting the volume removed in improvement fellings with the idea of sustaining a yield of low-grade timber. The rate of removal is determined by the volume of low-grade timber that can be sold, and by the funds available for investment in improving the growing stock. One improvement felling may be sufficient to free the better trees, but a few light fellings may give better results than one heavy felling. If gaps are created which are sufficiently large to permit restocking by natural or artificial regeneration, improvement fellings may be combined with enrichment (page 230).

Application in tropical mixed forest

Hutchinson (1987) has advocated the use of improvement fellings to favour the advanced growth of principal species that commonly appears in tropical mixed dipterocarpous forest in Sarawak following selective logging. Overmature and moribund trees of the old crop and weed species are removed to provide growing space for potentially useful saplings and young trees of principal species and prepare the stand for treatment under a silvicultural system.

Hutchinson based his proposals on several considerations. The first is classification of species into those that have already become accepted as principal species, those that are likely to become principal species in time, and the remainder. In Sarawak there is also a special category of species that are protected by law and cannot be felled. All this information provides the basis for a list of species to be favoured in improvement fellings.

The second consideration is classification of these species by their light-demanding and shade-bearing character. Whitmore (1984) has shown that in tropical rain forests of the Far East these two broad categories do not embrace all the variations that exist. He recognized four groups of species based on their increasing dependence on gaps in the canopy:

(1) those whose seedlings both establish and grow up under high forest;
(2) those that establish and grow up under high forest but show signs of benefiting from gaps;
(3) those that establish mainly in high forest but definitely require gaps to grow up; and
(4) those, the pioneer species, that establish mainly in gaps and only grow up in gaps.

Next comes consideration of suitable silvicultural treatment. If a conventional thinning is used to remove over-mature and competing trees so as to favour selected ones, the canopy is opened up abruptly creating large, well-lit gaps, physical damage is done to the growing stock as they fall, and Whitmore's last two categories of trees are favoured. If poison-girdling is used to remove unwanted trees the canopy opens gradually and to a limited extent. Less physical damage is done, Whitmore's first two categories are favoured, and at the same time the growth of woody climbers, coppice shoots, and light-demanding pioneer species is reduced.

In the mixed dipterocarpous forests of Sarawak selective logging during the period 1974 to 1980 was often light. Commonly, 5 to 15 trees producing $10-50\,\text{m}^3$ were harvested per hectare and the stands often contain promising advance growth. The actual process of improvement felling begins with enumerations in promising stands to identify 'leading desirables' together with an assessment of the condition of seedlings and saplings of the chosen list of species. The potential crop trees are stable, strong, and healthy. The smaller trees have intact leading shoots and the stems of the larger trees are damaged relatively little so that height and diameter growth will continue. Stem diameters range between 10 and 60 cm. As selection of potential crop trees proceeds the condition of the whole stand becomes evident and the appropriate silvicultural treatment is decided.

Enrichment

Fr. Enrichessement; Span. Enriquecimiento

The term enrichment embraces various measures for raising the proportion of principal species in degraded forests. In the tropics this is generally done by sowing or planting these in lines or strips or, less often, in gaps—either natural or resulting from exploitation. In temperate forests enrichment is often done by setting young trees of principal species at wide spacing among existing forest growth, which may be natural or planted. Like improvement fellings, enrichment is a preparatory stage in the progress toward a full silvicultural system. The term includes planting on land occupied by scrub.

Enrichment of tropical forest

The enrichment of moist tropical semi-deciduous or deciduous forest may be accomplished in lines cleared at intervals equal to or greater than the estimated mean diameter of the crowns of final crop trees; by this means the canopy will close before the trees are mature but intermediate yields of timber will be small or nil. This method has been used since the 1930s in several African countries (Lamb 1969; Kio and Ekwebelan 1987; Nwoboshi 1987). To be successful in the tropics planting in widely spaced lines must satisfy all of the following conditions:

1. The method is best applied where large timber and veneer logs are required. If smaller produce is essential the lines must be more closely spaced and clear felling followed by replanting will often be the best solution (see page 234).
2. The species planted must be fast growing (1.5 m or more per year is the minimum), naturally straight-stemmed, and self-pruning. These requirements are met by light-demanding species which colonize clear felled sites or fill gaps created in the canopy.
3. There must be no upper canopy. Only lines that are cleared by felling or poisoning or are situated in low secondary forest are suitable.
4. The regrowth between lines must not be easily set on fire.
5. Browsing animals must be absent, scarce, or of negligible effect on the planted trees.

The first step is to choose the species and the desired stem diameter at maturity. The corresponding mean crown diameter can be obtained by direct observation in the forest or by consulting published data relating crown diameter to stem diameter (see Dawkins 1959, 1963). The lines of trees are spaced at distances equal to or up to 20 per cent more than the estimated final mean crown diameter. This prevents competition between the crowns of trees in adjacent lines and provides some space for the development of natural seedlings of desirable species which may arise between the lines.

The spacing of trees within the lines is usually one-fifth of that between lines. This permits selection of one tree from four for the final crop. In lightly felled natural forest an overwood of trees which have been killed by herbicide will often be present and allowance must be made for losses among the new trees of up to 30 per cent. The spacing in the lines is therefore reduced to one-sixth or one-seventh of that between lines so as to ensure a final crop of well-formed trees.

The lines are cleared to a width of at least 2 m and movement along them for planting, tending, and protection is made easier by removing snags from woody growth along at least one side. Planting must follow immediately. Ideally poisoning of the upper canopy should also be timed so as progressively to let in light from the time of planting. The plants must also establish and grow rapidly. Pit planting is favoured and for many species use is made of containerized stock and less commonly of stumped plants or striplings (saplings 1.5 to 2 m tall from which all but the top two to four side branches and leaves have been removed). More rarely, direct sowing has been used, an example being *Cedrela odorata* in Africa.

Trees of secondary or weed species arising naturally between the lines must be cut or killed with herbicide before they actually overtop the planted trees. In Africa *Musanga cecropioides* and species of *Trema* and *Macaranga* are the most troublesome. Climbers over-arching from the regrowth beside the lines must also be cut back before they overshadow the new trees or impede access to the lines. Up to six or seven cleanings will be necessary during the first 12 months. As a working rule, the area enriched each year should not exceed one-eighth of the total area to be treated. Natural regeneration of secondary species such as *Alstonia congensis, Celtis soyauxii,* and *Ricinodendron africanum* is favoured as long as it does not endanger the primary species.

A thinning is usually required at the third or fourth year when the new trees should be well above the shrubs and climbers. It consists of selecting trees with straight stems and superior height growth; unless the variation in size is excessive these characters are more important than stem diameter. About half the crop will be removed to leave 40 to 50 trees to grow on to maturity.

The species most suited to planting in widely spaced lines in Africa are *Terminalia ivorensis* (in the wetter types of forest), *T. superba* (in less humid types), and *Triplochiton scleroxylon*. All these are gregarious, light-demanding species with straight stems and wide crowns. *Maesopsis eminii* also produces straight stems and wide crowns and reaches large sizes when planted in widely spaced lines in Uganda.

The Malaysian species *Shorea leprosula* and *S. parvifolia* are two suitable species with medium rates of growth. *Mansonia altissima, Tarrieta utilis, Entandophragma utile* and *E. cylindricum,* and *Lovoa trichiliodes* are species that have been tried in widely spaced line plantings because of their good form and valuable timbers, but their early growth is rather slow. Species of the *Meliaceae,* including *Chlorophora excelsa* and *Khaya ivorensis* have given disappointing results in west and east Africa because of slow and erratic early growth and attacks by the shoot borer *Hypsipyla robusta.* However line plantings of *Swietenia macrophylla* have succeeded in Fiji where the shoot borer does not occur.

Enrichment of temperate forest

The principles used in enriching degraded tropical and temperate forests are similar although the conditions for success are more stringent in the tropics.

In Britain there are two main approaches to enrichment (Evans 1984). The first is to accept much of the existing crop and restock gaps or clearings with the same species or a more productive species, provenance, or cultivar. Size of clearing must be related to the probable height of the better surrounding trees and should always be sufficient to allow several new trees to grow unimpeded, so that one or more will reach maturity. Thus if the better surrounding trees are expected to reach heights of 8 m a suitable minimum diameter is 15 to 17 m, giving an area of 0.02 ha (see page 179, and Appendix 1). Several clearings should be linked by access paths for tending purposes.

The second approach uses strips cleared at intervals, the remaining forest being left to maintain ground cover and a forested appearance. Up to half the area is restocked with one or more principal species. If the uncleared forest contains vigorous trees their crowns will expand rapidly and shade the new trees, so the width of cleared strip must be related to the final height of the best trees in the uncleared forest. If these are likely to attain 10 m a suitable minimum width of cleared strip is 19 to 21 m, sufficient for six to seven rows of plants. In general it has been found that enrichment using strips is easier to manage and usually cheaper than using groups.

Advantages and disadvantages

The advantages are:

1. In both temperate and tropical forests enrichment is useful where natural regeneration of principal species is deficient and cannot be induced in sufficient quantities and distribution because seed bearers are lacking or ecological conditions are unfavourable.
2. It provides a means of partial regeneration to increase the proportion of principal species from which natural regeneration may be expected in the future.
3. The genetic quality of the new crop can be improved by planting superior provenances and cultivars.
4. Forest conditions are retained and only a proportion of the area is disturbed or altered.
5. The work of clearing, planting, tending, and thinning is confined to definite lines, strips, or clearings so that it can be systematized and readily controlled.
6. Management of the forest is simplified and timber production can be concentrated where it is needed.
7. Yields of timber may be raised considerably in quantity and quality.

The disadvantages include:

1. Enrichment requires vigorous, sturdy, and uniform planting stock and the silvicultural operations must be thoroughly done and well-timed. For all these reasons skilled labour and supervision must be readily available.
2. The damage done by browsing animals can be severe because of the combination of easy access to the young trees and abundant dense cover.
3. The total costs per unit fully established depend on the state of the existing forest and speedy establishment. Enrichment can be expensive.

Application in practice

Enrichment of degraded forest has become more practicable because of several technical developments. Cultivars of many tropical and temperate species are now readily available and these can be planted at relatively wide spacing. Tools and machines continue to be developed for preparing lines, strips, and groups for planting and for tending the new crop. Chemical herbicides can also be used before and after planting to reduce competition from weeds and control encroachment by the surrounding trees and shrubs after planting. Effective and cheap tree shelters have been devised to enhance early growth of the new trees, allow use of herbicides close to them, and to provide protection from certain browsing animals.

In the tropics progress in enrichment has often been hampered by lack of training facilities and supervision to instil effective and safe working methods. In both temperate and tropical forestry enrichment has given variable results but is nevertheless worth thorough trial because of the very large areas of degraded forest that requires conversion to a productive condition. In the Solomon Islands very strong growth of climbers has made enrichment impracticable.

Replacement

When the existing forest has become so degraded that it contains a few trees of principal species (the basal area being around 4 m^2/ha) or the species remaining do not grow to a sufficient height to form high forest, replacement is usually the sole means of bringing it to a satisfactory state.

Replacement following clear felling is used where:

(1) the new crop will include light-demanding species;
(2) agricultural land is scarce and *taungya* is desirable (see page 79);

(3) intermediate yields of produce are required making thinnings desirable;

(4) coppice shoots are old and enfeebled and regrowth is unlikely to compete strongly with the new crop.

Sometimes the existing cover is cleared so that very persistent weed species can be destroyed by uprooting the stumps or applying herbicide. This is the case with *Rhododendron ponticum* in Britain which is very persistent under most tree crops.

Replacement under a shelterwood

One of the best methods of converting degraded forest into productive high forest is by artificial regeneration at regular spacing under the shelter of a light overwood. It is most readily used when much of the existing growth has reached the pole stage and the canopy is sufficiently high to allow easy access. At this stage the best stems are removed to provide income leaving an overwood of light-crowned trees. For best results the overwood should be uniformly stocked but must not be too dense. The temptation to retain good poles of desirable species to grow on with the new crop should normally be resisted because it is rarely successful.

When the new crop has become established the overwood is removed as soon as possible. This applies especially to species that benefit from some shade in early youth but are basically light-demanding, an example being Douglas fir. Removal of the overwood may be done by felling or herbicide; the former is preferred in places where amenity is important.

The advantages are:

1. Ground preparation costs are less than for complete clearance.
2. Planting costs are reduced and tender species are protected from frost, cold wind, and strong sunlight.
3. Soil water relations are better than on completely cleared sites. On heavy clays the site does not become too wet and weed growth is reduced.
4. The overwood is an advantage where amenity and field sports are important.

The disadvantages are:

1. Root competition and shade from the overwood can retard the growth of the new crop and care is needed to balance the growth of the two.
2. Very dry sites cannot support both the overwood and the new crop.
3. The cost of removing the overwood increases total establishment costs.

In Britain during the 1950s when the task of restoring forests made derelict during two world wars was tackled, much use was made of birch as an overwood to establish beech on thin chalk downland soils in southern England, and to establish Douglas fir, Western red cedar, and Silver firs on frosty sites in Wales and Scotland.

Conversion of coppice with standards to uniform broad-leaved high forest

Fr. Conversion en futaie pleine

This method called '*conversion classique*', with a preparatory period of rest, has been widely used in the state forests of France since the latter part of the nineteenth century. It leads up to the shelterwood uniform system and involves:

(1) a preparatory period of ageing or weakening of the coppice before an attempt is made at regeneration;

(2) accumulation of the largest possible number of standards of the younger age classes as seed bearers;

(3) covering the soil and gradually removing the cover.

The details vary according to local conditions, but an example will show how the conversion is usually made.

We may assume that the rotation for the principal species in the future high forest crop is fixed to correspond with an exploitable stem diameter, and is 120 years—which may conveniently divide into four periods of 30 years. The whole area may be divided into four corresponding periodic blocks; but in practice it is preferable to select only one block at a time, the crops allotted to this block being those containing most standards suitable as future seed bearers. So, one-quarter of the area is chosen as block I and it is no longer coppiced but is left for a preparatory period of 30 years (*période d'attente*) during which the coppice grows up and is either subjected to periodic light thinnings (*coupes préparatoires*) or left unthinned. Coppicing continues over the remainder of the forest but a larger number than usual of the smaller standards are reserved to aid future conversion. At the end of this preparatory period, regeneration fellings (*coupes de conversion*) are made in block I by opening the canopy gradually in a series of fellings. Restocking by natural regeneration, with as much supplementary artificial regeneration as needed, is completed within the 30 years.

During the period in which block I is being regenerated, periodic block

Table 3. Stages in converting coppice with standards to high forest.

Periods	Operations necessary in periodic blocks			
	I	II	III	IV
1961–1990	Select block I and allow preparatory period of rest. Thin coppice	Coppice cut as usual; these blocks not yet selected		
1991–2020	Regeneration fellings	Select block II and allow preparatory period of rest. Thin coppice	Coppice cut as usual; these blocks not yet selected	
2021–2050	Cleanings and thinnings	Regeneration fellings	Select block III and allow preparatory period of rest. Thin coppice	Coppice cut as usual
2051–2080	Thinnings	Cleanings and thinnings	Regeneration fellings	Allot block IV and allow preparatory period of rest. Thin coppice
2081–2110	Thinnings	Thinnings	Cleanings and thinnings	Regeneration fellings

II, also representing one-quarter of the total area, is lightly thinned for 30 years, the coppice being cut as usual over the remaining half of the forest. This procedure continues for periodic blocks III and IV in turn until the whole forest has been converted. Table 3 summarizes the steps, commencing for illustrative purposes in 1961.

The thinnings in the block allotted to the preparatory period of rest remove the large veterans (*anciens*) but reserve as many younger standards as possible to provide a full complement of seed bearers and shade out the coppice. Wherever the cover of the standards is insufficient to kill the coppice the latter is thinned periodically at intervals of 6 to 10 years, the first thinning being done during the year when the coppice should ordinarily be cut. The objectives of these thinnings are:

(1)   to furnish revenue from coppice shoots and any standards that may die off;

(2)   to remove coppice shoots interfering with the crowns of standards that will be future seed bearers; and

(3)   to reduce the number of stems progressively at each thinning until there are only two or three, or even one, left on each stool. This helps to kill out the coppice shoots, since shoots produced after such thinnings soon languish under the rest of the crop and the stools lose their vitality.

The regeneration fellings consist of the usual seeding, secondary, and final fellings characteristic of the uniform system. Standards are depended on as far as possible for seed bearers, because most coppice shoots, particularly of oak and beech, are not old enough to provide seed in quantity; however, some coppice shoots may have to be used as seed bearers. The seeding and secondary fellings are made cautiously to prevent vigorous regrowth of coppice. In the seeding fellings the canopy is raised by removing coppice shoots giving low cover; large standards, which are the best seed bearers are carefully retained. Groups of advance growth are encouraged by removing the cover from over and around them. The gradual opening of the canopy generally encourages regeneration of beech at the expense of more light-demanding species such as oak, and it may be necessary in places to cut the beech back.

Tending operations in regenerated crops consist of cleanings in the younger crops and thinnings later. Cleanings have to be done with great care so as to assist seedling regeneration against regrowth of coppice shoots.

Application in practice

The 'classical method' of conversion is applied today in forests that retain the structure of coppice with standards but the basal area is low, often in the range of 7–13 m²/ha. Enrichment and partial replacement are used with the classical method where necessary. The main disadvantage is the absolute need for continuous management extending over a century and a half, hence it is mainly seen in State rather than in privately owned forests.

Conversion by intensive reservation

Fr. Balivage intensif

A modification of the older system of conversion was devised by Aubert (1920) and has been developed into its modern form by Hubert (1979,

1981); with the object of increasing early returns in produce and money and shortening the time taken to complete the whole scheme of conversion.

Each annual coppice coupe, as its turn comes round for the normal coppice felling, is treated under intensive reservation or '*balivage intensif*', that is to say, at least 300 and usually more of the best standards and the best coppice shoots per hectare are retained to form the future high forest crop. As far as possible the older standards (most of the veterans and badly shaped first class standards) are removed and the smaller ones (tellers and the better second class standards) of oak, beech, and Wild cherry are retained and high pruned; coppice shoots, if retained, are thinned to one per stool. Subsequent operations consist of thinnings at intervals of 6 to 10 years beginning about 20 years later, followed subsequently by regeneration fellings when the coppice crop is not less than 60 years old, that is, not less than 30 years after the intensive reservation. As a rule, however, the average age of the crop is a good deal higher, since care is taken to select areas for regeneration that contain a good proportion of the larger standards for seeding purposes.

Periodic blocks are not fixed ahead. Areas for conversion to high forest are selected after careful enumeration of the crop together with soil capability classification. If there are many standards of the right sizes conversion may begin, but if not the crop is kept under coppice with standards and the number of standards retained is gradually increased so that a large number is accumulated. Crops under conversion generally contain many coppice shoots isolated and allowed to grow up, the normal density of the crop being thus secured; such coppice shoots may in time assist in providing seed for regeneration.

The chief difference between intensive reservation and the older classical method is that in the former, thinning of the coppice is done drastically at the start of the conversion period, whereas in the latter it takes place gradually throughout the period. Under *balivage intensif*, thinning the coppice to one shoot per stool gives the crop a somewhat open appearance but the canopy closes up satisfactorily after about 10 to 15 years.

Application in practice

On good soils this system has worked well, provided there are ample numbers of tellers and second class standards available for retention; where many first class and older standards are retained the results have not been so good, owing to the amount of space taken up by these large trees.

20

Agro-forestry systems

General description

The practice of growing timber trees, fruit trees, shrubs, and palms in combination with agricultural crops and animals is very widespread and takes many forms. Forestry may be combined with agriculture to improve the microclimate for field crops and animals, maintain soil fertility, control erosion, produce fuelwood and timber, and increase the cash income from the land. In 1978 the International Council for Research in Agro-forestry (ICRAF), sponsored by Canada, Holland, and Switzerland set up its headquarters in Kenya 'to improve the nutritional, economic, and social well-being of developing countries by the promotion of agro-forestry systems designed to result in better land use without detriment to the environment' (ICRAF 1983).

ICRAF has made an inventory of existing practices and systems around the world. The classification recognizes agro-silvicultural systems in which agricultural crops are grown with trees; silvo-pastoral systems where the trees provide fodder and shaded pasture for grazing animals; agro-silvo-pastoral systems which combine all three; and 'home gardens' such as those in the humid parts of southern Asia which have many-storied mixtures of trees, shrubs, and other plants, akin to the structure of tropical moist forest on a very small scale.

Two of the systems described in this book can be classified as agro-silvicultural systems and two others are used as silvicultural components of agro-forestry systems. The purpose of this chapter is to draw attention to these silvicultural systems and indicate the contributions that silvicultural systems in general can make to agro-forestry systems.

Agro-silvicultural systems

Two systems in this category are described in detail. These are:

(1) Clear cutting using artificial regeneration with the aid of field crops (see pages 79 to 81); and
(2) Coppice with field crops (see pages 201 to 202).

In China, since 1958, forestry has become closely integrated with agriculture through the concept of 'four around forestry' in support of agriculture (Kemp 1980). Trees are planted along roadsides, on the banks of rivers and canals, around fields, and near houses and villages as part of integrated rural development to strengthen the economy of rural communes. During the first 2 or 3 years after planting much of the land between the rows of trees is intercropped with cereals, vegetables, fodder crops, and green manure. The land is kept clear of weeds, fire hazard is greatly reduced, the trees establish well, and significant amounts of food are produced.

These agro-silvicultural systems become true systems only when the supply of goods and services by the two components can be sustained. Thus crops of trees intended to support agriculture by providing shelter from wind must be established, tended, and regenerated in such a way that they will continue to perform their function effectively. It is rarely, if ever, sufficient to plant a shelterbelt which is then retained untended until it is over-mature or diseased. A growing stock that is suitably constituted in age or size classes is essential and there must also be sufficient flexibility in design and the allocation of land to allow for changes in agricultural practices. The landscape of many parts of Europe, including the British Isles, is full of old, decaying tree crops, originally intended as shelterbelts, which no longer perform their original function. Because they often have high amenity and conservation values they also create conflict because they impede the introduction of improved or different agricultural practices. By contrast the collective shelterstrips used in the Jutland region of Denmark are being modified to take account of new knowledge which improves their effectiveness in supporting agriculture and incidentally enhances the beauty of the landscape.

Two other silvicultural systems are often used as components in agro-forestry systems:

(1) Pollarding (see page 208);
(2) Simple coppice, usually on short rotations (see pages 190 and 195).

It is also tempting to use the coppice with standards system but for the reasons presented on pages 222 to 224, simple coppice and the clear cutting system hold more promise for tropical countries.

Silvo-pastoral systems

In New Zealand Radiata pine has been grown at wide spacing on pasture for about 15 years. A thorough programme of research from 1971 to 1974 identified several conditions essential for success (Levack 1986, p. 72). There is a choice between agro-forestry or a gradual transition from farming to forestry. So far farmers are more interested in the former and foresters in the latter.

The trees are genetically improved seedlings or rooted cuttings planted in groups or rows which are widely spaced. The number of final crop trees lies between 100 and 200/ha. Closer spacing shades the ground cover and generates too much slash. The trees are protected from excessive damage by the grazing animals by several managerial practices, such as timing the introduction of animals, attending to details of where the trees are planted, and the use of electrified fencing. Frequent pruning is done to leave 3–4 m of live crown whilst keeping the knotty core to specified dimensions and maximizing the volume of clear timber. Early thinning within the tree crop is essential so that defective individuals are removed quickly.

Turning to grazing in forests (Davis 1976) has described the usual situation succinctly. Grazing of domestic livestock and timber production are of limited compatibility as land uses, mainly because grass and trees are often vigorous competitors for the same land and one will dominate the other. 'A land use choice is necessary: either emphasize grazing use and retain some trees for protective livestock cover or emphasize tree growth and sharply restrict or exclude grazing.'

Provided that the operation is carefully planned and controlled, limited grazing of livestock in forests can sometimes be accommodated with advantage in the clear cutting system. An example from New Zealand demonstrates the technique (Levack 1986). After clear felling and replanting the new crop of Radiata pine, forage species are sown and the coupe is fenced. When the young crop is established livestock are introduced. Their grazing can be beneficial in improving access to the stand, controlling weeds, reducing the risk of fire, and stimulating tree growth through recycling of nutrients through the livestock. Access to areas of open pasture is needed if continuous grazing is not desirable. Reference to the description of grazing by deer in the Austrian mountains (page 45) will demonstrate one origin of these ideas.

Agro-silvo-pastoral systems

Beaton (1987) described a successful system from the west midlands of England which begins as a form of agro-silviculture and becomes, with the passage of time, a silvo-pastoral system.

Two-year-old rooted plants, or unrooted sets 2 m tall, of several cultivars of poplar were planted deeply into moist, fertile but heavy soil at a spacing of 8 m—with the object of producing trees of 45 cm diameter at breast height in 22 years. The planting pattern was triangular and the young trees were positioned accurately to facilitate later cultivation of field crops and control of weeds by machines. Repeated pruning was done to minimize the knotty core and maximize the volume of clear timber. By the ninth year when pruning was complete, each tree had about 6 m of clear stem. Epicormic shoots were removed annually from all the trees.

Poplars are intolerant of competition from grasses and to permit rapid early growth the ground was kept clean by cultivation. When the trees were established the agricultural component was introduced. The triangular spacing gave about 6.5 m between the rows of trees. A strip of 1 m was left uncropped on either side of the rows giving bands of 4.5 m for field crops. The regime that was evolved consisted of cereals or other field crops in alternate bands, the remaining bands being kept fallow. This gave a two-course rotation of field crops and fallow, and produced a workable and profitable compromise between the requirements of adequate increment from the trees and a satisfactory yield from the field crops.

The maximum useful period of cropping was around 8 years. By this time the trees had reached heights of 10 m with live crowns 4.5–5.5 m long. The ground was then sown down to a permanent pasture of grass and clover (*Trifolium* sp.) and grazing units of 12–24 ha were formed for sheep and cattle. To forestall damage to roots by soil compaction and to the stems by debarking, grazing was restricted to frost-free months, that is, from April to October. The value of the grazing declined after 5 to 6 years as the quantity of leaf fall increased. By this time the poplar trees were close to maturity and ripe for felling.

References

Allison, B. J. (1980). Tomorrow's plantations today. In *Forest plantations: the shape of the future.* Weyerhaeuser Science Symposia, No. 1, pp. 7–23, Tacoma.

Ammon, W. (1951). *Das plenterprinzip in der waldwirtschaft* (3rd edn). Haupt, Bern.

Anderson, M. L. (1949). Some observations on Belgian forestry. *Empire Forestry Review*, **28**, 117–30.

—— (1960). Norway spruce – Silver fir – beech mixed selection forest. *Scottish Forestry*, **14**, 87–93.

Anon. (1948). *Coppicing and pollarding of Cassia siamea.* Report of the Forest Department of Tanganyika, 1945. Dar es Salaam.

Anon. (1950). Service forestier du Soudan: rapport pour l'année 1949. *Bois et Forêts Tropiques*, **15**, 294–5.

Arnborg, T. (1985). Planting of trees and shrubs for protection of land and for production of fodder and fuelwood. In *Forest energy and the fuelwood crisis,* Report, No. 41 (eds. G. Siren and C. P. Mitchell), pp. 93–111. Swedish University of Agricultural Sciences, Uppsala.

Assmann, E. (1970). *The principles of forest yield study* (ed. P. W. Davis). Pergamon, Oxford.

Attiwill, P. M. (1979). Nutrient cycling in *Eucalyptus obliqua* (L'Hérit.) forest. III Growth, biomass, and net primary production. *Australian Journal of Botany*, **27**, 439–58.

Aubert, C. G. (1920). La conversion des taillis en futaie dans l'ouest de la France. *Revue des Eaux et Forêts*, **58**, 153–60, 189–214, 227–34.

Auclair, D. (1982). Present and future management of coppice in France. In *Broadleaves in Britain* (eds. D. C. Malcolm, J. Evans, and P. N. Edwards), pp. 40–6. Institute of Chartered Foresters, Edinburgh.

Baidoe, J. F. (1970). The selection system as practised in Ghana. *Commonwealth Forestry Review*, **49**, 159–65.

Ballard, R. and Gessel, S. P. (ed.) (1983). *IUFRO Symposium on forest site and continuous productivity,* General Technical Report, PNW-163. Pacific Northwest Forest and Range Experiment Station, Portland.

Banti, G. (1948). A system of harvesting *Robinia* practised in the Lombard bush country. *Italia Forestale e Montana*, **3**, 41–5.

Barnard, R. C. (1955). Silviculture in the tropical rain forest of western Nigeria compared with Malayan methods. *Empire Forestry Review*, **34**, 355–68.

Barnes, R. D. and Gibson, G. L. (1984). *Provenance and genetic improvement strategies in tropical forest trees.* Commonwealth Forestry Institute, University of Oxford, and Zimbabwe Forestry Commission, Harare.

Barrett, J. W., Martin, R. E., and Wood, D. C. (1983). Northwestern Ponderosa pine and associated species. In *Silvicultural systems for the major forest types of the United States*. Agriculture Handbook, No. 445 (ed. R. M. Burns), pp. 16–18. US Department of Agriculture, Forest Service, Washington.

Barrette, B. R. (1966). *Redwood (Sequoia sempervirens) sprouts in Jackson state forest*, State Forest Note, No. 29. California Division of Forestry, Sacramento.

Bates, C. G., Hilton, H. C., and Krueger, T. (1929). Experiments in the silvicultural control of natural regeneration of Lodgepole pine in the central Rocky Mountains. *Journal of Agricultural Research*, **38**, 229–43.

Bauer, F. (ed.) (1986). *Diagnosis and classification of new types of damage affecting forests* (special edn). Commission of the European Communities, Brussels.

Beaton, A. (1987). Poplars and agroforestry. *Quarterly Journal of Forestry*, **81**, 225–33.

Beck, D. E. and Sims, D. H. (1983). Yellow-poplar. In *Silvicultural systems for the major forest types of the United States*, Agriculture Handbook, No. 445 (ed. R. M. Burns), pp. 180–2. US Department of Agriculture, Forest Service, Washington.

Bevan, D. (1962). The Ambrosia beetle or Pinhole borer *Trypodendron lineatum*. *Scottish Forestry*, **16**, 94–9.

—— (1987). *Forest insects*, Forestry Commission Handbook, No. 1. HMSO, London.

Billany, D. J. (1978). *Gilpinia hercyniae—a pest of spruce*, Forestry Commission Forest Record, No. 117. HMSO, London.

Biolley, H. (1920). *L'aménagement des forêts par la méthode éxperimentale et specialement la méthode du contrôle*. Attinger Frères, Paris.

Blackburn, P., Petty, J. A., and Miller, K. F. (1988). An assessment of the static and dynamic factors involved in windthrow. *Forestry*, **61**, 29–43.

Blyth, J. (1986). *Edinburgh University experimental area, Glentress forest*, Internal Report. Department of Forestry and Natural Resources, University of Edinburgh.

Bolton, Lord. (1956). *Profitable forestry*, pp. 99–106. Faber and Faber, London.

Booth, T. C. (1974). Topography and wind risk. *Irish Forestry*, **31**, 130–4.

—— (1977). *Windthrow hazard classification*. Forestry Commission Research Information Note, No. 22/77/SILN. Farnham.

—— (1984). Natural regeneration in the native pinewoods of Scotland. *Scottish Forestry*, **38**, 33–42.

Bormann, F. H., Likens, G. E., Siccama, T. G., Pierce, R. S., and Eaton, J. S. (1974). The effect of deforestation on ecosystem export and the steady state condition at Hubbard Brook. *Ecological Monographs*, **44**, 255–77.

Bouvarel, P. (1981). The outlook for energy forestry in France and in the European Community. In *Energy from biomass* (eds. W. Palz, P. Chartier, and D. O. Hall), pp. 172–80. Applied Science Publications, London.

Boyd, J. M. (1987). Commercial forests and woods: the nature conservation baseline. *Forestry*, **60**, 113–34.

Bradford, Lord. (1981). An experiment in irregular forestry. *Y Coedwigwr*, **33**, 26–30.

Brazier, J. D. (1977). The effect of forest practices on the quality of the harvested crop. *Forestry*, **50**, 49–66.

Brown, A. H. F. (1986). The effects of tree mixtures on soil and tree growth. In *Discussion on uneven-aged silviculture* (ed. C. Cahalan), pp. 32–40. Department of Forestry and Wood Science, University College of North Wales, Bangor.

Brown, J. M. B. (1953). *Studies on British beechwoods*, Forestry Commission Bulletin, No. 20, pp. 38–9. HMSO, London.

Brünig, E. F. (1973). Storm damage as a risk factor in wood production in the most important wood-producing regions of the earth. *Forstarchiv*, **44**, 137–40. (Translation 4339, W. Linnard. Commonwealth Forestry Bureau, Oxford.)

―――― (1980). The means to excellence through control of growing stock. In *Forest plantations: the shape of the future*, Weyerhaeuser Science Symposium, No. 1, pp. 201–24. Weyerhaeuser, Tacoma.

Brundtland, G. H. (1987). In *Our common future*, pp. ix–xv. The World Commission on Environment and Development, Oxford University Press.

Buffet, M. (1980). *La régénération du Chêne rouvre*, Bulletin Technique, No. 12, pp. 3–29. Office National des Forêts, Paris.

―――― (1981). *La régénération du Hêtre en plaine*, Bulletin Technique, No. 13, pp. 29–47. Office National des Forêts, Paris.

Burley, J. and Plumptre, R. A. (1985). Species selection and breeding for fuelwood plantations. In *Forest energy and the fuelwood crisis*, Report, No. 41 (eds. G. Siren and C. P. Mitchell), pp. 42–58. Swedish Academy of Agricultural Sciences, Uppsala.

Burns, R. M. (ed.) (1983). Silvicultural systems for the major forest types of the United States Agriculture handbook. No. 445. US Department of Agriculture, Forest Service, Washington.

Cadman, W. A. (1965). The developing forest as a habitat for animals and birds. *Forestry*, **38**, 168–72.

Cannell, M. G. R. (1979). Biological opportunities for genetic improvement in forest productivity. In *The ecology of even-aged forest plantations* (eds. E. D. Ford, D. C. Malcolm, and J. Atterson), pp. 119–44. NERC Institute of Terrestrial Ecology, Cambridge.

―――― (1980). Productivity of closely spaced young poplar on agricultural soils in Britain. *Forestry*, **53**, 1–21.

Champion, H. G., Seth, H. K., and Khattak, G. M. (1973). *Manual of general silviculture for Pakistan*. Pakistan Forest Institute, Peshawar.

Chard, J. S. R. (1966). The Red deer of Furness fells. *Forestry*, **39**, 135–50.

―――― (1970). *The Roe deer*, Forestry Commission Leaflet, No. 45. HMSO, London.

Conway, S. (1982). *Logging practices* (revised edn.). Miller-Freeman, San Francisco.

Cotta, H. von (1817). *Anweisung zum waldbau*. Dresden.

Crooke, M. (1979). The development of populations of insects. In *The ecology of even-aged forest plantations* (eds. E. D. Ford, D. C. Malcolm, and J. Atterson), pp. 209–17. NERC Institute of Terrestrial Ecology, Cambridge.

Crossley, D. I. (1956). *Effect of crown cover and slash density on the release of seed from slash-borne Lodgepole pine cones*, Forest Research Division, Technical Note, No. 41. Canada Department of Northern Affairs and Natural Resources, Ottawa.

Crowe, S. (1978). *The landscape of forests and woods*, Forestry Commission Booklet, No. 44. HMSO, London.

Crowther, R. E. (1976). *Guidelines to forest weed control*, Forestry Commission Leaflet, No. 66. HMSO, London.

Crowther, R. E. and Evans, J. (1984). *Coppice*, Forestry Commission Leaflet, No. 83. HMSO, London.

Coutts, M. P. (1986). Components of tree stability in Sitka spruce on peaty gley soil. *Forestry*, **59**, 173–97.

Coutts, M. P. and Philipson, J. J. (1987). Structure and physiology of Sitka spruce roots. *Proceedings Royal Society of Edinburgh*, **93B**, 131–44.

Curdel, R. (1973). Les taillis et taillis-sous-futaie du Nord-Pas-de-Calais et de la Picardie. Evolution, production et culture. *Revue Forestière Française*, **25**, 413–24.

Davis, K. P. (1976). *Land use*. McGraw-Hill, New York.

Davis, L. S. and Johnson, K. (1987). *Forest management* (3rd edn.). McGraw-Hill, New York.

Dawkins, H. C. (1959). The volume increment of natural tropical high forest and limitations on its improvement. *Empire Forestry Review*, **38**, 175–80.

—— (1963). Crown diameters: their relation to bole diameter in tropical forest trees. *Commonwealth Forestry Review*, **42**, 318–33.

Delabraze, P. (1986). Sylviculture Méditerranéene. In *Précis de sylviculture* (ed. L. Lanier), pp. 362–94. École Nationale du Genie Rural, des Eaux et des Forêts, Nancy.

Dieterich, H., Müller, S., and Schlenker, G. (1970). *Urwald von morgen*. Ulmer, Stuttgart.

Directorate of Forests (1974). *Working plans for the Chittagong tracts, North Forest Division*. Bangladesh Government, Dacca.

Dubordieu, J. (1986*a*). Sylviculture en montagne. In *Précis de sylviculture* (ed. L. Lanier), pp. 349–61. École Nationale du Genie Rural, des Eaux et des Forêts, Nancy.

—— (1986*b*). Notions d'aménagement des forêts. In *Précis de sylviculture* (ed. L. Lanier), pp. 395–408. École Nationale du Genie Rural, des Eaux et des Forêts, Nancy.

Eberhard, J. (1922). Der schirmkeilschlag und die Langenbrander wirtschaft. *Forstwissenschaftliches Centralblatt*, **66**, 41–76.

Edie, A. G. (1916). Thinnings of teak coppice in the pole areas of Kanara. *Indian Forester*, **42**, 157–9.

Edwards, I. D. (1981). The conservation of the Glen Tanar native pinewood, near Aboyne, Aberdeenshire. *Scottish Forestry*, **35**, 173–8.

Eifert, J. (1903). Forstliche sturm-beobachtungen im Mittelgebirge. *Allgemeine Forst-und Jagzeitung*, **79**, 323–83.

Ek, A. R., Balsiger, J. W., Biging, G. S., and Payenden, B. (1976). A model for determining optimal clear-cut strip width for Black spruce harvest and re-generation. *Canadian Journal of Forest Research*, **6**, 382–8.

El Atta, H. A. and Hayes, A. J. (1987). Decay in Norway spruce caused by *Stereum sanguinolentum* Alb. and Schw. ex Fr., developing from extraction wounds. *Forestry*, **60**, 101–11.

Evans, J. (1982). *Plantation forestry in the tropics*. Oxford University Press.

—— (1984). *Silviculture of broadleaved woodland*, Forestry Commission Bulletin, No. 62, pp. 26–9. HMSO, London.

—— (1987). Tree shelters. In *Advances in practical arboriculture* (ed. D. Patch), Forestry Commission Bulletin, No. 65, pp. 67–76. HMSO, London.

Evans, H. F. and King, C. J. (1988). *Dendroctonus micans: guidelines for forest managers*, Research Note, No. 128. Forestry Commission, Farnham.

Farrell, P. W., Flinn, D. W., Squire, R. O., and Craig, F. G. (1983). On the maintenance of productivity of Radiata pine monocultures on sandy soils in south-east Australia. In *IUFRO Symposium on forest site and continuous productivity* (eds. R. Ballard and S. P. Gessel), General Technical Report, PNW–163, pp. 117–28. Pacific Northwest Forest and Range Experiment Station, Portland.

Faulkner, R. (1987). Genetics and breeding of Sitka spruce. *Proceedings Royal Society of Edinburgh*, **93B**, 41–50.

Fishwick, R. W. (1965). Neem plantations in northern Niger. In Replenishing the world's forests. *Commonwealth Forestry Review*, **62**, p. 205.

Flöhr, W. (1969). Results of the conversion of pure Scots pine stands into mixed forest by underplanting in the Sauen district. *Archiv für Forstwesen*, **18**, 991–4.

Flury, P. (1914). *La Suisse Forestière*. Payot, Lausanne.

Food and Agriculture Organisation (1979*a*). *Forestry for rural communities*. FAO, Rome.

—— (1979*b*). *Economic analysis of forestry projects*, Forestry Paper, No. 17, Supplement No. 1. FAO, Rome.

——(1980). *Poplars and willows in wood production and land use*, Forestry Series, No. 1. FAO, Rome.

Ford, E. D., Malcolm, D. C., and Atterson, J. (eds.) (1979). *The ecology of even-aged plantations*. NERC Institute of Terrestrial Ecology, Cambridge.

Forestry Commission (1983). *Census of woodlands and trees 1979–1982: England*. Forestry Commission, Edinburgh.

Fourchy, P. (1954). Some aspects of present-day silviculture in Switzerland. *Quarterly Journal of Forestry*, **48**, 85–104.

Fowells, H. A. (ed.) (1965). *Silvics of forest trees of the United States*, Agriculture Handbook, No. 271. US Department of Agriculture, Forest Service, Washington.

Garcia, O. (1986). Forest estate modelling (Part 2). In *1986 Forestry handbook* (ed. H. Levack), pp. 97–9. New Zealand Institute of Foresters, Wellington North.

Garfitt, J. E. (1979). The importance of brashing. *Quarterly Journal of Forestry*, **73**, 153–4.

—— (1980). Treatment of natural regeneration and young broadleaved crops. *Quarterly Journal of Forestry*, **74**, 236–9.

—— (1984). The group selection system. *Quarterly Journal of Forestry*, **78**, 155–8.

Gayer, K. (1880). *Der waldbau*. Parey, Berlin.
—— (1886). *Der gemischtewald*. Parey, Berlin.
Ghosh, R. C. (1983). In Replenishing the world's forests. *Commonwealth Forestry Review*, **62**, p. 205.
Gilliusson, R. (1985). Wood energy: global needs. In *Forest energy and the fuelwood crisis*, Report. No. 41 (eds. G. Siren and C. P. Mitchell), pp. 2–15. Swedsh Academy of Agricultural Sciences, Uppsala.
Glew, D. R. (1963). *The results of stand treatment in the White spruce-Alpine fir type of the northern interior of British Columbia*, Forest Management Notes, No. 1. British Columbia Forest Service, Victoria.
Godwin, G. E. and Boyd, J. M. (1976). The Black Wood of Rannoch—a new forest nature reserve. *Scottish Forestry*, **30**, 192–4.
Gregory, S. C. and Redfern, D. B. (1987). The pathology of Sitka spruce in northern Britain. *Proceedings Royal Society of Edinburgh*, **93B**, 145–56.
Greig, B. J. W. (1981). *Decay fungi in conifers*, Forestry Commission Leaflet, No. 79. HMSO, London.
Greig, B. J. W. and Low, J. D. (1975). An experiment to control *Fomes annosus* in second rotation pine crops. *Forestry*, **48**, 147–63.
Greig, B. J. W. and Redfern, D. B. (1974). *Fomes annosus*, Forestry Commission Leaflet, No. 5. HMSO, London.
Gysel, L. W. (1957). Effects of silvicultural practices on wildlife food and cover in oak and aspen types in northern Michigan. *Journal of Forestry*, **55**, 803–9.
Hagner, S. (1962). *Natural regeneration under shelterwood stands*. Meddelanden från Statens Skogsforskningsinstitut, 52. Stockholm.
Hartig, G. L. (1791). *Anweisung zur holzzucht für förster*. Marburg.
Haufe, H. (1952). *Thirty years of the blendersaumschlag system in Württemburg: its results in practice*. Sauerlander, Frankfurt-am-Main.
Helliwell, D. R. (1982). *Options in forestry*. Packard, Chichester.
Hibberd, B. G. (1985). Restructuring of plantations in Kielder forest district. *Forestry*, **58**, 119–29.
—— (ed.) (1986). *Forestry practice*, Forestry Commission Bulletin, No. 14, pp. 68–75. HMSO, London.
Hiley, W. E. (1959). Two-storied high forest. *Forestry*, **32**, 113–16.
——(1967). *Woodland management* (2nd edn), p. 307. Faber and Faber, London.
Hill, H. W. (1979). Severe damage to forests in Canterbury, New Zealand resulting from orographically re-inforced winds. In *Symposium on forest meteorology*, World Meteorological Organization, Report No. 527, pp. 22–40. Canadian Forest Service, Ottawa.
Holmsgaard, E., Holstener-Jørgensen, H., and Yde-Andersen, A. (1961). Soil formation, increment, and health of first and second generation stands of Norway spruce I. Nordsjaelland. *Forstlige Forsogsvaesen i Danmark*, **27**, 1–167.
Holtam, B. W. (ed.) (1971). *Windblow of Scottish forests in January 1968*, Forestry Commission Bulletin, No. 45. HMSO, London.
Howell, B. N., Harley, R. M., White, R. D. F., and Lamb, R. G. M. (1983). The Dartington story II. *Quarterly Journal of Forestry*, **77**, 5–16.

Howland, P. (1969). Effects of singling coppice in *Eucalyptus saligna* wood fuel crops at Muguga, Kenya. *East African Agricultural and Forestry Journal*, **35**, 66–7.

Hubert, M. (1979). *Le balivage: une solution économique et sans risque pour la mise en valeur de certains taillis ou taillis-sous-futaie pauvres.* Institut pour le Développement Forestier, Paris.

―― (1981). *Du taillis a la futaie: 8 Arguments en faveur du balivage*, Forêts de France et Action Forestière, No. 244, pp. 13–17.

Hutchinson, I. D. (1987). Improvement thinning in natural tropical forest. In *Natural management of tropical moist forest* (eds. F. Mergen and J. R. Vincent), pp. 113–34. Yale University, New Haven.

Hutt, P. and Watkins, K. (1971). The Bradford plan for continuous forest cover. *Journal of the Devon Trust for Nature Conservation*, **3**, 69–74.

Hütte, P. (1968). Experiments on windflow and wind damage in Germany; site and susceptibility of spruce forests to storm damage. In *Wind effects on the forest*. Supplement to *Forestry*, pp. 20–6.

Innes, J. L. (1987). *Air pollution and forestry*, Forestry Commission Bulletin, No. 70. HMSO, London.

International Council for Research in Agroforestry (1983). A global study of agroforestry systems: a project announcement. *Agroforestry Systems*, **1**, 169–73.

Jacobs, M. R. (ed.) (1980). *Eucalypts for planting* (2nd edn.), Forestry Series, No. 11. FAO, Rome.

James, N. D. G. (1982). *The forester's companion* (3rd edn.), pp. 60–1. Blackwell, Oxford.

Jenkins, D. and Reusz, H. III. Prinz. (1969). A successful case history reconciling forestry and Red deer management. *Forestry*, **42**, 21–7.

Jezewski, Z. (1959). The effect of different methods of osier plantation management on root development and yields. *Sylwan*, **103**, 61–71.

Jobling, J. and Pearce, M. L. (1977). *Free growth of oak*, Forestry Commission Forest Record, No. 113. HMSO, London.

Jones, A. B. (1984). Forestry equipment for export. In *International aspects of forestry* (eds. J. G. S. Gill and D. C. Malcolm), pp. 51–60. Institute of Chartered Foresters, Edinburgh.

Jones, A. P. D. (1948). *The natural forest inviolate plot*. Nigerian Forest Department, Zaria.

Jones, E. W. (1947). Scots pine regeneration in a New Forest inclosure. *Forestry*, **21**, 151–78.

―― (1965). Pure conifers in Europe. *Journal of the Oxford University Forestry Society*, **13**, 3–15.

Johnson, J. von (1984). Prescribed burning: requiem or renaissance? *Journal of Forestry*, **82**, 82–91.

Johnston, D. R. (1978). Irregularity in British forestry. *Forestry*, **57**, 163–9.

Johnston, D. R., Grayson, A. J., and Bradley, R. T. (1967). *Forest planning*. Faber and Faber, London.

Kemp, R. H. (1980). Forestry in China: II, a Commonwealth connection. *Commonwealth Forestry Review*, **59**, 53–60.

Khattak, G. M. (1976). History of forest management in Pakistan, III, irrigated plantations and riverain forests. *Pakistan Journal of Forestry*, **26**, 231–41.

Kio, P. R. O. and Ekwebelan, S. A. (1987). Plantations versus natural forests for meeting Nigeria's wood needs. In *Natural management of tropical moist forests* (eds. F. Mergen and J. R. Vincent), pp. 149–76. Yale University, New Haven.

Knüchel, H. (1953). *Planning and control in the managed forest* (trans. M. L. Anderson). Oliver and Boyd, Edinburgh.

König, E. and Gossow, H. (1979). Even-aged stands as habitat for deer in central Europe. In *The ecology of even-aged forest plantations* (eds. E. D. Ford, D. C. Malcolm, and J. Atterson), pp. 429–51. NERC Institute of Terrestrial Ecology, Cambridge.

Köstler, J. (1956). *Silviculture* (trans. M. L. Anderson). Oliver and Boyd, Edinburgh.

Laacke, R. J. and Fiske, J. N. (1983). Sierra Nevada mixed conifers. In *Silvicultural systems for the major forest types of the United States*, Agriculture Handbook, No. 445 (ed. R. M. Burns), pp. 44–7. US Department of Agriculture, Forest Service, Washington.

Lamb, A. F. A. (1969). Artificial regeneration within the humid tropical forest. *Commonwealth Forestry Review*, **48**, 41–53.

Lanier, L. (1986). *Précis de sylviculture*. École Nationale du Genie Rural, des Eaux et des Forêts, Nancy.

Larsen, C. S. (1937). *The employment of species, types, and individuals in forestry*. Royal Veterinary and Agricultural College Yearbook, Reitzel, Copenhagen.

Laurie, M. V. (1974). *Tree planting practices in African savannas*. Forestry Development Paper, No. 19. FAO, Rome.

Leakey, R. R. B. (1987). Clonal forestry in the tropics—a review of developments, strategies, and opportunities. *Commonwealth Forestry Review*, **66**, 61–75.

Leibundgut, H. (1984). *Die waldpflege* (3rd edn). Paul Haupt, Bern.

Levack, H. (ed.) (1986). *1986 forestry handbook*. New Zealand Institute of Foresters, Wellington North.

Liddon, E. M. and Hector, N. (1984). Letter, *Management of Basket willows*.

Lin, D. Y. (1956). China. In *A world geography of forest resources* (eds. S. Haden-Guest, J. K. Wright, and E. M. Teclaff), pp. 529–50. Ronald Press, New York.

Liocourt, F. de (1898). De l'aménagement des sapinières. *Bulletin de la Société forestière de Franche-Comté et du Territoire de Belfort*, **4**, 396–409, 645–7.

Low, A. J. (ed.) (1985). *Guide to upland restocking practice*, Forestry Commission Leaflet, No. 84. HMSO, London.

———— (1988). Scarification as an aid to natural regeneration in the Glen Tanar native pinewood. *Scottish Forestry*, **42**, 15–20.

Maitre, H. F. (1987). Natural forest management in Côte d'Ivoire. *Unasylva*, **39**, 53–60.

Malcolm, D. C. and Studholme, W. P. (1972). Yield and form in high elevation stands of Sitka spruce and European larch in Scotland. *Scottish Forestry*, **26**, 298–308.

Mantel, K. (1964). History of the international science of forestry, with special reference to central Europe. *International Review of Forest Research*, **1**, 1–37.

Matte, V. H. (1965). Coppicing of *Pinus radiata*. *Naturwissenschaften*, **2**, 90–1.

Matthews, J. D. (1963). Factors affecting the production of seed by forest trees. *Forestry Abstracts*, **24**, 1–13.

—— (1964). Seed production and seed certification. *Unasylva*, **18**, 73–4, 104–18.

—— (1975). Prospects for improvement by site amelioration, breeding, and protection. *Philosophical Transactions Royal Society of London*, **B271**, 115–38.

McAlpine, R. G., Brown, C. L., Herrick, A. M., and Ruark, H. E. (1966). Silage sycamore. *Forest Farmer*, **26**, 6–7, 16.

McKell, C. M. and Finnis, J. M. (1957). Control of soil moisture depletion through use of 2,4-D on a mustard nurse crop during Douglas fir seedling establishment. *Forest Science*, **3**, 332–5.

McNeil, J. D. and Thompson, D. A. (1982). Natural regeneration of Sitka spruce in the forest of Ae. *Scottish Forestry*, **36**, 269–82.

Meiggs, R. (1982). *Trees and timber in the ancient Mediterranean world*. Oxford University Press.

Métro, A. (ed.) (1955). *Eucalypts for planting*, Forestry and Forest Products Studies, No. 11. FAO, Rome.

Miegroet, M. van (1962). The silvicultural treatment of small woodlands. *Bulletin de la Société Royale Forestière de Belgique*, **69**, 437–56.

Miller, H. G. (1981). Forest fertilisation: some guiding concepts. *Forestry*, **54**, 157–67.

—— (1983). *Wood energy plantations—diagnosis of nutrient deficiencies and the prescription of fertiliser applications in biomass production*, International Energy Agency—Forest Energy Agreement, Report, No. 3. Ministry of Natural Resources, Maple, Ontario.

—— (1984). Water in forests. *Scottish Forestry*, **38**, 165–81.

Miller, K. F. (1985). *Windthrow hazard classification*, Forestry Commission Leaflet, No. 85. HMSO, London.

——, Quine, C. P., and Hunt, J. (1987). The assessment of wind exposure for forestry in Great Britain. *Forestry*, **60**, 179–92.

Mills, D. H. (1980). *The management of forest streams*, Forestry Commission Leaflet, No. 78. HMSO, London.

Mitchell, C. P. and Puccioni-Agnoletti, M. C. (eds.) (1983). *Forest energy plantations on forest sites*, International Energy Agency Report, No. NE 1983:1. National Swedish Board for Energy Source Development, Stockholm.

Muhl, R. G. (1987). Contour felling for cable crane and forwarder extraction on steep slopes. *Commonwealth Forestry Review*, **66**, 273–80.

Munger, T. T. (1940). The cycle from Douglas fir to hemlock. *Ecology*, **21**, 451–9.

Murphy, G. (1982). Directional felling of old crop *Pinus radiata* on steep country. *New Zealand Journal of Forestry*, **27**, 67–76.

Murray, J. S. (1979). The development of populations of pests and pathogens in even-aged plantations—fungi. In *The ecology of even-aged forest plantations* (eds. E. D. Ford, D. C. Malcolm, and J. Atterson), pp. 193–208. NERC Institute of Terrestrial Ecology, Cambridge.

Murray, J. S. and Young, C. W. T. (1961). *Group dying of conifers*, Forestry Commission Forest Record, No. 46. HMSO, London.

National Academy of Sciences (1980). *Firewood crops. Shrub and tree species for energy production.* Vol. 1. National Academy Press, Washington.

—— (1981). *Sowing forests from the air.* National Academy Press, Washington.

Neustein, S. A. (1965). Windthrow on the margins of various sizes of felling area. In *Forestry Commission report on forest research 1964*, pp. 166–71. HMSO, London.

Nutter, W. L. (1979). Effects of forest plantations on the quantity, quality, and timing of water supplies. In *The ecology of even-aged forest plantations* (eds. E. D. Ford, D. C. Malcolm, and J. Atterson), pp. 351–67. NERC Institute of Terrestrial Ecology, Cambridge.

Nwoboshi, L. C. (1987). Regeneration success of natural management, enrichment planting, and plantations of native species in west Africa. In *Natural management of tropical moist forests* (eds. F. Mergen and J. R. Vincent), pp. 72–92. Yale University, New Haven.

O'Carroll, N. (1978). The nursing of Sitka spruce, 1. Japanese larch. *Irish Forestry*, **35**, 60–5.

Oliver, W. W., Powers, R. F., and Fiske, J. N. (1983). Pacific Ponderosa pine. In *Silvicultural systems for the major forest types of the United States*, Agriculture Handbook, No. 445 (ed. R. M. Burns), pp. 48–52. US Department of Agriculture, Forest Service, Washington.

O'Loughlin, C. (1986). Forestry and hydrology. In *1986 forestry handbook* (ed. H. Levack), pp. 13–15. New Zealand Institute of Foresters, Wellington North.

Osmaston, F. C. (1968). *The management of forests.* Allen and Unwin, London.

Oswald, H. (1982). Silviculture of oak and beech high forests in France. In *Broadleaves in Britain* (eds. D. C. Malcolm, J. Evans, and P. N. Edwards), pp. 31–9. Institute of Chartered Foresters, Edinburgh.

Pawsey, R. G. and Gladman, R. J. (1965). *Decay in standing conifers developing from extraction damage*, Forestry Commission Forest Record, No. 64. HMSO, London.

Peace, T. R. (1961). The dangerous concept of the natural forest. *Quarterly Journal of Forestry*, **55**, 12–23.

—— (1962). *Pathology of trees and shrubs, with special reference to Britain.* Oxford University Press.

Penistan, M. J. (1960). Forestry in the Belgian uplands. *Forestry*, **33**, 1–7.

Peterken, G. F. (1977). General management principles for nature conservation in British woodlands. *Forestry*, **50**, 27–48.

Petty, J. A. and Swain, C. (1985). Factors influencing breakage of conifers in high winds. *Forestry*, **58**, 75–84.

Petty, J. A. and Worrell, R. (1981). Stability of coniferous tree stems in relation to damage by snow. *Forestry*, **54**, 115–28.

Philip, M. S. (1983). *Measuring trees and forests*. University of Dar es Salaam, Tanzania.

—— (1986). *Management systems in the tropical moist forests of Africa*. FAO, Rome.

Philipp, K. (1926). *Die umstellung der wirtschaft in der Badischen staats-, gemeinde- und körperschaftswaldungen*. Lang, Karlsruhe.

Plaisance, G. (1966). A successful conversion, the work of B. Lorentz 1775–1865. *Revue Forestière Française*, **18**, 82–98.

Pratt, J. E. (1979). *Fomes annosus* butt rot of Sitka spruce. *Forestry*, **52**, 11–45, 113–27.

Pyatt, D. G. (1970). *Soil groups of upland forests*, Forestry Commission Forest Record, No. 71. HMSO, London.

Rackham, O. (1980). *Ancient woodland—its history, vegetation, and uses in England*. Edward Arnold, London.

Ratcliffe, P. R. (1985). *Glades for deer control in upland forests*, Forestry Commission Leaflet, No. 86. HMSO, London.

Reade, M. G. (1965). Natural regeneration of beech. *Quarterly Journal of Forestry*, **59**, 121–31.

Robertson, F. C. (1971). *Terminology of forest science, technology, practice, and products*. Society of American Foresters, Washington.

Rochelle, J. A. and Bunnell, F. L. (1979). Plantation management and vertebrate wildlife. In *The ecology of even-aged forest plantations* (eds. E. D. Ford, D. C. Malcolm, and J. Atterson), pp. 389–411. NERC Institute of Terrestrial Ecology, Cambridge.

Roisin, P. (1959). La transformation des pessières. *Bulletin de la Société Royale Forestière de Belgique*, **66**, 153–90.

Rollinson, T. J. D. and Evans, J. (1987). *The yield of Sweet chestnut coppice*, Forestry Commission Bulletin, No. 64. HMSO, London.

Ronco, F. Jr. and Ready, K. L. (1983). Southwestern Ponderosa pine. In *Silvicultural systems for the major forest types of the United States*, Agricultural Handbook, No. 445 (ed. R. M. Burns), pp. 70–2. US Department of Agriculture, Forest Service, Washington.

Rouse, G. D. (1984). Spain, 1981. *Quarterly Journal of Forestry*, **78**, 104–11.

Rowan, A. A. (1976). *Forest road planning*. Forestry Commission Booklet, No. 43. HMSO, London.

—— (1977). *Terrain classification*, Forestry Commission Forest Record, No. 114. HMSO, London.

Ryker, R. A. and Lowsensky, J. (1983). Ponderosa pine and Rocky Mountain Douglas fir. In *Silvicultural systems for the major forest types of the United States*, Agricultural Handbook, No. 445 (ed. R. M. Burns), pp. 53–5. US Department of Agriculture, Forest Service, Washington.

Sahlen, K. (1984). Reforestation results after direct sowing under plastic cones. *Sveriges Skogsvårdsförbunds Tidskrift*, **82**, 20–45.

Samapuddhi, K. (1974). Thailand's forest villages. *Unasylva*, **27**, 20–3.

Sanchez, P. (1976). *Properties and management of soils in the tropics*. Wiley, New York.

Savill, P. S. (1983). Silviculture in windy climates. *Forestry Abstracts*, **44**, 473–88.

Savill, P. and Evans, J. (1986). *Plantation silviculture in temperate regions*. Oxford University Press.

Schädelin, W. (1937). *L'éclaircie. Traitement des forêts par la selection qualitative* (trans. M. Droz). Attinger, Paris.

Schmidt, R. (1987). Tropical rain forest management—a status report. *Unasylva*, **39**, 2–17.

Science Council of Canada. (1973). *A national statement by the schools of forestry at Canadian universities.* Ottawa.

Scott, T. M. (1972). *The Pine shoot moth and related species*, Forestry Commission Forest Record, No. 83. HMSO, London.

Scott, T. M. and King, C. J. (1974). *The Large pine weevil and Black pine beetles*, Forestry Commission Leaflet, No. 58. HMSO, London.

Sharp, L. (1975). Timber, science, and economic reform in the seventeenth century *Forestry*, **48**, 51–86.

Shoulders, E. and Parham, G. (1983). Slash pine. In *Silvicultural systems for the major forest types of the United States*. Agriculture Handbook, No. 445 (ed. R. M. Burns), pp. 162–6. US Department of Agriculture, Forest Service, Washington.

Silversides, C. R. (1981). Innovative transportation in the 2000's. In *Forest to mill: challenges of the future*, Weyerhaeuser Science Symposium, No. 3. Weyerhaeuser, Tacoma.

Smith, D. M. (1986). *The practice of silviculture* (8th edn). Wiley, New York.

Society of American Foresters (1984). Prescribed burning. *Journal of Forestry*, **82**, 82–91.

Solbrana, K. (1982). *Preliminary results from funnel sowing of conifers*, Rapport, No. 4/ 82. Norsk Institutt for Skogsforskning, Ås.

Somerville, A. (1980). Wind stability: forest layout and silviculture. *New Zealand Journal of Forestry Science*, **10**, 476–501.

Spears, J. S. (1983). Replenishing the world's forests: tropical reforestation; an achievable goal? *Commonwealth Forestry Review*, **62**, 201–17.

Spurr, S. H. and Barnes, B. V. (1980). *Forest ecology* (3rd edn), pp. 421–57. Wiley, Chichester.

Squire, R. O. (1983). Review of second rotation silviculture of *Pinus radiata* plantations in southern Australia: establishment practices and expectations. *Australian Forestry*, **46**, 83–90.

Squire, R. O. and Flinn, D. W. (1981). *Site disturbance and nutrient economy of plantations with special reference to Radiata pine on sands*, Proceedings Australian Forest Nutrition Workshop, pp. 291–302. CSIRO, Division of Forest Research, Canberra.

Steele, R. C. and Peterken, G. F. (1982). Management objectives for broadleaved woodland conservation. In *Broadleaves in Britain* (eds. D. C. Malcolm, J. Evans, and P. N. Edwards), pp. 91–103. Institute of Chartered Foresters, Edinburgh.

Steven, H. M. and Carlisle, A. C. (1959). *The native pinewoods of Scotland*. Oliver and Boyd, Edinburgh.

Steward, P. J. (1980). Coppice with standards: a system for the future. *Commonwealth Forestry Review*, **59**, 149–54.

Stone, E. L. (1975). Effects of species on nutrient cycles and soil change. *Philosophical Transactions of the Royal Society of London*, **B271**, 149–62.

Stott, K. G. (1956). Cultivation and uses of Basket willows. *Quarterly Journal of Forestry*, **50**, 103–12.

Susmel, L. (1986). Prodromi di una nuova selvicoltura. *Annali, Accademia Italiana di Scienze Forestali*, **35**, 33–51.

Sutton, R. F. (1969). *Form and development of conifer root systems*, Commonwealth Forestry Bureau, Technical Communication No. 7. Commonwealth Agricultural Bureaux, Farnham Royal.

Sweet, G. B. (1975). Flowering and seed production. In *Seed orchards* (ed. R. Faulkner), pp. 72–82. Forestry Commission Bulletin, No. 54. HMSO, London.

Tackle, D. (1954). *Lodgepole pine management in the intermountain region: a problem analysis*, Intermountain Forest and Range Experiment Station, Publication, No. 2. US Department of Agriculture, Forest Service, Ogden, Utah.

Tassy, L. (1872). *Études sur l'aménagement des forêts* (2nd edn), pp. 165–72. Rothschild, Paris.

Taylor, C. M. A. (1985). The return of nursing mixtures. *Forestry and British Timber*, **14**, 18–19.

Timber Growers United Kingdom (1985). *The forestry and woodland code*. TGUK Ltd, London.

Thomasius, H., Butter, D., and Marsch, M. (1986). *Massnahmen zur stabilisierung von fichtenforsten gegenüber schnee- und sturmschäden*. Proceedings, 18th World Congress of the International Union of Forestry Research Organisations, pp. 1–53. Ljubjana.

Thompson, D. A. (1984). *Ploughing of forest soils*, Forestry Commission Leaflet, No. 71. HMSO, London.

Tompa, K. (1963). *Experiments for determining suitable spacing for osiers*, Erdészeti és Faipari Egytem, No. 1–2, pp. 147–57. Sopron.

Troup, R. S. (1928). *Silvicultural systems*. Oxford University Press.

—— (1952). *Silvicultural systems* (2nd edn), (ed. E. W. Jones). Oxford University Press.

Tubbs, C. H., Jacobs, R. D., and Cutler, D. (1983). Northern hardwoods. In *Silvicultural systems for the major forest types of the United States*, Agriculture Handbook, No. 445 (ed. R. M. Burns), pp. 121–7. US Department of Agriculture, Forest Service, Washington.

Turner, G. (1959). Note relative à la transformation des pessières en station. *Bulletin de la Société Royale Forestière de Belgique*, **66**, 414–20.

Wagner, C. (1912). *Die blendersaumschlag und sein system*. Laupp, Tubingen.

—— (1923). *Die grundlagen der raumlichen ordnung in wald* (4th edn). Laupp, Tubingen.

Watt, A. S. (1919). On the causes of failure of natural regeneration in British oakwoods. *Journal of Ecology*, **7**, 173–203.

Watt, A. S. (1923). On the ecology of British beechwoods with special reference to their regeneration, Part I. The causes of the failure of natural regeneration of beech. *Journal of Ecology*, **11**, 1–48.

Weidemann, E. (1923). *Zuwachsrückgang und wuchsstockungen bei der fichte im den mittleren und den unteren hohenlagen der Sachsischen stadtsforsten.* Laux, Tharandt.

—— (1924). *Fichtenwachtum und humuszustand.* Arbeiten aus der Biologischen Reichanstalt fur Land- und Forstwirtschaft, No. 13, pp, 1–177.

Welch, D., Chambers, M. G., Scott, D., and Staines, B. W. (1988). Roe-deer browsing on spring-flush growth of Sitka spruce. *Scottish Forestry*, **42**, 33–43.

Wesseley, J. (1853). *Die Österreichischen alpenlander und ihre forsten*, Vol. 1, p. 300. Vienna.

Whitehead, D. (1982). Ecological aspects of natural and plantation forests. *Forestry Abstracts*, **43**, 615–24.

Whitmore, T. C. (1984). *Tropical rain forests of the Far East* (2nd edn). Oxford University Press.

Williamson, R. L. and Twombly, A. D. (1983). Pacific Douglas fir. In *Silvicultural systems for the major forest types of the United States*, Agriculture Handbook, No. 445 (ed. R. M. Burns), pp. 9–13. US Department of Agriculture, Forest Service, Washington.

Winterflood, E. G. (1976). The forests of East Anglia. *Forestry*, **49**, 23–8.

Wood, R. F., Miller, A. D. S., and Nimmo, M. (1967). *Experiments on the rehabilitation of uneconomic broadleaved woodlands*, Forestry Commission Research and Development Paper, No. 51. HMSO, London.

Wyatt-Smith, J. (1987). Problems and prospects for natural management of tropical moist forests. In *Natural management of tropical moist forests* (eds. F. Mergen, and J. R. Vincent), pp. 6–22. Yale University, New Haven.

Wyatt-Smith, J. and Panton, W. P. (1963). *Manual of Malayan silviculture for inland forests.* Malayan Forest Records, No. 23, III-4, pp. 1–13.

Zavitkowski, J. (1979). Energy production in irrigated, intensively cultured plantations of *Populus* 'Tristis' and Jack pine. *Forest Science*, **25**, 383–92.

Zobel, B. J. and Talbert, J. (1984). *Applied forest tree improvement.* Wiley, Chichester.

Zundel, R. (1960). *Yield studies in two-aged stands of Scots pine over Silver fir in north Württemberg*, Schriftenreihe der Land Landesforstverwaltung Baden-Württemberg, No. 6. Stuttgart.

Appendix 1.   List of plant names

Abies, Pinaceae
A. alba Mill.   European silver fir
A. grandis (Dougl.) Lindl.   Grand fir
A. concolor var. *lowiana* (Gord.) Lemm.   California white fir
Acacia, Leguminosae/Mimosoideae
A. modesta Wall.
A. saligna (Labill.) H. Wendl.
A. senegal (L.) Willd.   Gum arabic
Acer, Aceraceae
A. campestre L.   Field maple
A. macrophyllum Pursh   Bigleaf maple
A. pseudoplatanus L.   sycamore
Ailanthus altissima (Mill.) Swingle, Simaroubaceae   Tree of Heaven
Aira flexuosa L., Gramineae   Wavy hair grass
Albizia lebbek (L.) Benth., Leguminosae/Mimosoideae kokko
Alnus, Betulaceae
A. incana (L.) Moench   Grey alder
A. rubra Bong.   Red alder
Alstonia congensis Engl., Apocynaceae   alstonia
Anemone nemorosa L., Ranunculaceae   Wood anemone
Aningeria robusta Aubrev and Pellegr., Sapotaceae   aningeria
Arceuthobium species, Loranthaceae   Dwarf mistletoe
Azadirachta indica A. Juss., Meliaceae   neem
Betula, Betulaceae
B. papyrifera Marsh.   Paper birch
B. pendula Roth.   Silver birch
B. pubescens Ehrh.   Downy birch
Brassica juncea (L.) Czern., Cruciferae   mustard
Calliandra calothyrsus Meissn., Leguminosae/   calliandra
Mimosoideae
Calluna vulgaris (L.) Hull, Ericaceae   heather
Carpinus betulus L., Betulaceae   hornbeam

Carya species, Juglandaceae	hickory
Cassia siamea Lam., Leguminosae/Caesalpinoideae	Yellow cassia
Castanea sativa Mill., Fagaceae	Sweet chestnut
Cedrela odorata L., Meliaceae	
Celtis soyauxii Engl., Ulmaceae	
Chamaecyparis, Cupressaceae	
C. nootkatensis (D. Don) Spach	Alaska yellow cedar
C. obtusa (Sieb. and Zucc.) Endl.	hinoki
Chlorophora excelsa (Welw.) Benth. and Hook. f.	iroko
Cirsium arvense (L.) Scop., Compositae	Creeping thistle
Clerodendron species, Verbenaceae	
Convolvulus arvensis L., Convolvulaceae	Lesser bindweed
Corylus avellana L., Betulaceae	hazel
Cryptomeria japonica (L.f.) Don, Taxodiaceae	Japanese red cedar
Cunninghamia lanceolata (Lamb.) Hook. f., Taxodiaceae	Chinese fir
Cynometra alexandri C. H. Wright, Leguminosae/ Caesalpinoideae	muhimbi
Dalbergia sissoo Roxb., Leguminosae/Papilionaceae	sissoo
Deschampsia caespitosa (L.) Beauv., Gramineae	Tufted hair grass
Digitalis purpurea L., Scrophulariaceae	foxglove
Dioscorea species, Dioscoreaceae	yams
Diospyros mespiliformis Hoecht. ex A. DC., Ebenaceae	ebony
Dipterocarpus species, Dipterocarpaceae	dipterocarps
Dyera costulata (Miq.) Hook. f., Apocynaceae	jelutong
Endospermum malaccense M.A., Euphorbiaceae	
Entandrophragma, Meliaceae	
E. angolense (Welw.) C.DC.	gedu nohor
E. cylindricum (Sprague) Sprague	sapele
E. utile (Dawe and Sprague) Sprague	utile
Erythrophleum suavolens (Gull. and Parr.) Brenan, Leguminosae/Caesalpinoideae	missanda
Eucalyptus, Myrtaceae	
E. camaldulensis Dehnh.	River red gum
E. cloeziana F. Muell.	Gympie messmate
E. delegatensis R.T. Bak.	Tasmanian oak
E. globulus Labill.	Blue gum
E. gomphocephala A.DC.	tuart

E. grandis W. Hill ex Maiden	Flooded gum
E. microtheca F. Muell.	coolibah
E. paniculata Sm.	Grey ironbark
E. regnans F. Muell.	Mountain ash
E. saligna J. E. Smith	Sydney blue gum
E. tereticornis J. E. Smith	Forest red gum
Eugenia jambos L., synonym *Jambosa jambos* (L.) Millsp., Myrtaceae	
Fagus sylvatica L., Fagaceae	beech
Ficus species, Moraceae	
Fraxinus excelsior L., Oleaceae	ash
Gambeya delevoyi, synonym *Chrysophyllum delevoyi* De Wild Sapotaceae	
Gmelina arborea Roxb., Verbenaceae	yemane
Gonostylus bancanus (Miq.) Kurz, Thymelacaceae	ramin
Guarea cedrata (A. Chev.) Pellegr., Meliaceae	guarea
Hevea brasiliensis (Willd. ex A. Juss.) M.A., Euphorbiaceae	rubber
Juglans species, Juglandaceae	walnut
Juncus effusus L. Juncaceae	Soft rush
Khaya, Meliaceae	
K. anthotheca (Welw.) C.DC.	African mahogany
K. ivorensis A. Chev.	African mahogany
Lagerstroemia speciosa (L.) Pers., Lythraceae	pyinma
Lamium galeobdolon (L.) L., Labiatae	Yellow archangel
Larix, Pinaceae	
L. decidua Mill.	European larch
L. leptolepis (Sieb. and Zucc.) Endl.	Japanese larch
L. occidentalis Nutt.	Western larch
Leucaena leucocephala (Lam.) de Wit, Leguminosae/ Mimosoideae	leucaena
Libocedrus decurrens Torrey, Cupressaceae	Incense cedar
Liquidambar styraciflua L., Hamamelidaceae	Sweetgum
Liriodendron tulipifera L., Magnoliaceae	Yellow poplar
Lovoa, Meliaceae	
L. brownii Sprague	African walnut
L. trichilioides Harms.	African walnut

Luzula sylvatica (Huds.) Gaudin, Juncaceae	Great woodrush
Macaranga species, Euphorbiaceae	
Maesopsis eminii Engl., Rhamnaceae	musizi
Manihot esculenta Crantz, Euphorbiaceae	cassava
Mansonia altissima A. Chev., Sterculiaceae	mansonia
Mildbraediodendron excelsum Harms., Papilionaceae	
Molinia caerulea (L.) Moench, Gramineae	Purple moor grass
Morus nigra L., Moraceae	Black mulberry
Musanga cecropioides R.Br., Moraceae	
Nesogordonia papaverifera (A.Chev.) Capuron, Sterculiaceae	danta
Nothofagus, Fagaceae	
N. dombeyi Bl.	
N. procera (Poepp. and Endl.) Oerst.	
Olea cuspidata Well. Cat., Oleaceae	
Oxalis acetosella L., Oxalidiaceae	Wood sorrel
Pericopsis elata van Meeuwen, Sophoreae	afrormosia
Pentaspodon motleyi Hook. f., Anacardiaceae	
Picea, Pinaceae	
P. abies (L.) Karst.	Norway spruce
P. glauca (Moench) Voss	White spruce
P. mariana (Mill.) B.S.P.	Black spruce
P. sitchensis (Bong.) Carr.	Sitka spruce
Pinus, Pinaceae	
P. banksiana Lamb.	Jack pine
P. canariensis C. Smith	Canary Island pine
P. contorta Dougl. var. *contorta*	Coastal lodgepole pine
P. contorta Dougl. var. *latifolia* Wats.	Inland lodgepole pine
P. echinata Mill.	Shortleaf pine
P. elliottii Engl. var. *elliottii*	Slash pine
P. halepensis Mill.	Aleppo pine
P. lambertiana Dougl.	Sugar pine
P. nigra Arnold	Black pine
P. nigra var. *maritima* (Ait.) Melville	Corsican pine
P. nigra Arnold var. *nigra*	Austrian pine
P. oocarpa Scheide	Oocarpa pine

P. palustris Mill.	Longleaf pine
P. pinaster Ait.	Maritime pine
P. ponderosa Dougl.	Ponderosa pine
P. radiata D.Don	Radiata pine
P. resinosa Ait.	Red pine
P. rigida Mill.	Pitch pine
P. roxburghii Sargent	Chir pine
P. strobus L.	Eastern white pine
P. sylvestris L.	Scots pine
P. taeda L.	Loblolly pine
Platanus, Platanaceae	
P. occidentalis L.	American plane
P. × *hispanica* Muenchh.	London plane
Populus, Salicaceae	
P. alba L.	White poplar
P. canescens (Ait.) Sm.	Grey poplar
P. grandidentata Michx.	Bigtooth aspen
P. tremula L.	aspen
P. tremuloides Michx.	Quaking aspen
P. trichocarpa Torr. and Gray	Western balsam poplar
Prosopis juliflora (Swartz) DC., Leguminosae/ Mimosoideae	mesquite
Prunus avium L., Rosaceae	Wild cherry
Pseudotsuga menziesii (Mirb.) Franco, Pinaceae	Douglas fir
Pteris aquilinum (L.) Kuhn, Polypodiaceae	bracken
Quercus, Fagaceae	
Q. ilex L.	Holm oak
Q. kelloggii Newb.	California black oak
Q. petraea (Mattuschka) Liebl.	Sessile oak
Q. robur L.	Pedunculate oak
Ranunculus ficaria L., Ranunculaceae	Lesser celandine
Rhododendron ponticum L., Ericaceae	rhododendron
Ricinodendron africanum Muell. Arg., Euphorbiaceae	
Robinia pseudoacacia L., Leguminosae/Papilionaceae	Locust tree
Rubus, Rosaceae	
R. fruticosus agg.	blackberry-bramble

R. idaeus L.	raspberry
Rumex species, Polygonaceae	dock
Salix, Salicaceae	
S. alba L.	White willow
S. alba var. *vitellina* Stokes	
S. 'Americana'	American osier
S. caprea L.	Great sallow
S. gracilis Anderss.	
S. purpurea L.	Purple willow
S. rigida Muhl.	
S. triandra L.	Almond willow
S. viminalis L.	Common osier
Sarothamnus scoparius (L.) Wimm. ex Koch, Leguminosae/Papilionaceae	broom
Scottellia species, Flacourtiaceae	
Sequoia sempervirens (D.Don) Endl., Taxodiaceae	Coast redwood
Shorea, Dipterocarpaceae	
S. leprosula Miq.	
S. parvifolia Dyer	
S. robusta Gaertn.f.	sal
Sorbus aucuparia L., Rosaceae	rowan
Swietinea macrophylla King, Meliaceae	Central American mahogany
Tamarix species, Tamaricaceae	tamarisk
Tarrietia utilis (Sprague) Sprague, Sterculiaceae	niangon
Tectona grandis L.f., Verbenaceae	teak
Tephrosia candida (Roxb.) DC., Leguminosae/Papilionaceae	
Terminalia, Combretaceae	
T. ivorensis A.Chev.	idigbo
T. superba Engl. and Diels	afara, limba
Tetraclinis articulata (Vahl) Mast., Cupressaceae	
Teucrium scorodonia L., Labiatae	Wood sage
Thuja plicata D.Don, Cupressaceae	Western red cedar
Tilia cordata Mill., Tiliaceae	Small-leaved lime
Trema species, Ulmaceae	
Trifolium species, Leguminosae/Papilionaceae	clovers

Triplochiton scleroxylon K. Schum., Sterculiaceae	obeche
Tsuga heterophylla (Raf.) Sarg., Pinaceae	Western hemlock
Ulmus procera Salisb., Ulmaceae	English elm
Urtica dioica L., Urticaceae	Stinging nettle
Zea mays L., Gramineae	maize
Zizyphus mauritanica Lam., Rhamnaceae	Indian jujube

Appendix 2. Dimensions of clearings to provide growing space for trees of final size

Diameter of group (metres; yards)		Area (hectares; acres)		Number of plants (2 × 2 m spacing)	Number of trees of final size Douglas fir;	ash, sycamore;	oak
11–13	12–14	0.01	0.02	25– 35	3	1	1
15–17	16–18	0.02	0.05	45– 55	5	2	1
19–21	21–22	0.03	0.07	70– 85	8	4	2
22–24	23–25	0.04	0.10	95–115	11	5	2
25–26	27–28	0.05	0.12	120–130	14	6	3
27–28	29–30	0.06	0.15	145–155	16	7	3
29–30	32–33	0.07	0.17	165–175	18	8	4
31–32	34–35	0.08	0.20	180–200	21	9	4
33–34	36–37	0.09	0.22	215–230	24	10	5
35–36	38–39	0.10	0.25	240–255	27	12	5
37–38	40–41	0.11	0.27	270–285	30	13	6
39–40	42–44	0.12	0.30	300–315	33	15	6
41	45	0.13	0.32	330	36	16	7
42–43	46–47	0.14	0.36	345–365	38	17	7
44	48	0.15	0.37	380	42	19	8
45–46	49–50	0.16	0.39	400	44	20	9
47	51	0.17	0.42	435	48	21	9
48–49	52–53	0.18	0.44	450–470	50	22	10
50–51	54–56	0.20	0.50	490–510	54	24	15
62	68	0.30	0.75	755	83	37	20
72	79	0.40	1.00	1020	113	50	25
80	89	0.50	1.23	1255	139	62	30
88	96	0.60	1.48	1520	169	75	32
90	100	0.63	1.56	1590	177	78	

Index